How Things Don't Work

Other Books by the Authors

by James Hennessey and Victor Papanek:

Nomadic Furniture, Pantheon, 1973
Nomadic Furniture 2, Pantheon, 1974

by Victor Papanek:

Design for the Real World: Human Ecology and Social Change, Pantheon, 1971
"Big Character" Poster No. 1: Work Chart for Designers, Finn Sloth, Postbox 51,
Charlottenlund, Denmark, 1973

How Things Don't Work

Victor Papanek and James Hennessey

Photographs, illustrations, and designs by the Authors,
Sara Hennessey, Ira Velinsky, and Mike Whalley

 Pantheon Books, New York

All rights reserved under International and Pan-American Copyright Conventions. Published in the United States by Pantheon Books, a division of Random House, Inc., New York, and simultaneously in Canada by Random House of Canada Limited, Toronto.

Selection on page 117 from *Newsweek,* April 22, 1974 (International Edition). Copyright © 1974 by Newsweek, Inc. All rights reserved. Reprinted by permission.

Library of Congress Cataloging in Publication Data

Papanek, Victor J.
  How Things Don't Work.

  Bibliography: p. 149.
  1. Consumer education.   2.   Commercial products.
3. Design, Industrial.   I.   Hennessey, James, joint
author.   II.   Title.

TX335.P34   1977   640.73   76–62709
ISBN 0–394–49251–X
ISBN 0–394–73324–X pbk.

Designed by Irva Mandelbaum

Manufactured in the United States of America

First Edition

We dedicate this book to Michael Brian, Christopher John and Lizanne Kathrin Hennessey, and Jennifer Satu Papanek, in the hope that things will work for them and never the other way around.

Contents

List of Illustrations

Figure

Figure

Preface

Methane Indigestion, Solar Power and What *You* Can Do

. . . so by stating everything by reason, and by making the most rational judgment of things, every man may be in time master of every mechanic art. I had never handled a tool in my life, and yet in time, by labor, application, I found at last that I wanted nothing but could have made it, especially if I had had tools.

—Daniel Defoe, *Robinson Crusoe*

We must recognize the obvious. It costs more to produce our present forms of ugliness than to create better alternatives. We will be forced (like it or not) toward better, saner and more energy-saving tools and devices simply because we cannot afford any other kind. We will make things in a more decentralized and participatory way, in environments better suited to our minds, bodies and tasks. These alternatives are inescapable; they will lead us to elegant solutions: that is, making things we really need in apt ways.

Massive problems faced by workers and users need innovative remedies. Societies, and the individuals making up social groups, tend to respond in a number of different ways to each new problem. There is the capitalist approach (make it bigger), the technocratic one (make it better), the "revolutionary" solution (portray the problem as an example of an exploitative system) and the pre-industrial romantic fallacy (don't use it; maybe it will go away by itself). We propose a fifth alternative response: *Let's invent a different answer.*

If we take any one of the problems that beset us, such as "non-biodegradable plastics" or "The Automobile," and analyze it from these various possible viewpoints, we can get startlingly different responses. The capitalist will reply that the use of plastics or automobiles is inevitable, and that drawbacks that may exist are more than canceled by the many benefits. The technocrat will glibly explain away environmental hazards by claiming that they can be eliminated through a (still undetermined) better technology. The "revolutionary" answer will be that these are but the symptoms of a decadent capitalistic system, and can be eliminated only through socialist planning. The pre-industrial romantic will advocate immediately doing away with all plastics and automobiles. His answer to the question of pollution will be to close down all factories and promote a move back to the land.

Our alternative response is that we can use biodegradable materials that already exist, invent others and design nonpolluting technologies; that we can furthermore design technologies that are more compatible with life-styles fitted to human scale. In the meantime, those of us who have decided (or are forced) to live in a mass society based on a high technology must make the best of it. In this book we hope to give you ways of coping with life under these circumstances.

In our first chapter we attempt to use your bathtub—and the bathroom surrounding—as a simple case study of how a combination of things have been going wrong for a long time, leading to a distastrous collection of unworkable devices. We pose alternative solutions you can build or explore on your own.

The concept of *sharing rather than owning* things is developed in Chapter 2. Many products that we purchase at high initial outlay, and maintain expensively for years after, are underused. We develop alternative suggestions for the redesign and rebuilding of products that can be shared.

In Chapters 3 and 4 we investigate the possibility of people building things for themselves from plans or kits. We consider the social implications over a long-range period. We also try to take into account that people will gain a better understanding of products they themselves build and, therefore, will feel less alienated from them.

In our fifth chapter we develop several checklists to enable you to clearly determine whether a purchase that you might consider necessary is needed at all. We challenge the increasing use of electric and electronic devices that make products more expensive, and frequently less workable, and offer alternative suggestions.

In Chapter 6 we explore the ever-more-confusing dividing line between package and contents. We give examples of useless, expensive or dangerous overpackaging, and also explore the welter of nonsense and near-nonsense products that engulf us.

The public environment is our main concern in Chapter 7. Since such fields as city planning, regional planning, and architecture lie far beyond the scope of this book, we consider only those objects in the public environment to which people are forced to relate.

Chapter 8 attempts to provide you with the know-how and hence the confidence to buy things used. We offer hints, checklists and alternative suggestions to help you maintain and repair your present possessions and things bought second-hand. We provide another checklist that warns you away from things that it might be unwise to buy used.

The various ways in which obsolescence can be manipulated to make you spend more money or replace good things you now have are analyzed in Chapter 9.

In our final chapter we deal with consumer participation in the making of new products. We also attempt an honest overview of the immediate future and how this may affect your choices.

With these ten or more strategies we hope to help you cope with the technological age. If you are looking for radical answers to smash the present system, you will not find them here. To quote Hundertwasser in *Rainy Day*,

> People have no preparation for looking at what is beautiful. If you offer a glimpse of paradise before the revolution has triumphed, you are branded a traitor. Perhaps I'm a traitor when instead of going in for constant criticism or destruction, I try to do something constructive and to guide people to a world that is—well, just how I like to picture a better world.

This, then, is not a book for people who are into alternative-energy research. There is, however, a great deal of valuable study and development in this field going on all over the world. Tidal power is in use at St. Malo and on the Loire and Garonne rivers in France. There are villages entirely heated and lit by solar furnaces at Mt. Louis and Odeillo. The city of Gothenburg in Sweden has used methane power for nearly two generations. Windmills create power routinely in some of the Arab states and Israel, and major wind generators are being built in Canada as this is being written. In the Alps, near Montblanc, glacial power is being put to practical use, although only on a prototypal level so far. The first geothermal power plant was built at Ladarello in Italy more than seventy years ago. Since then, geothermal power has been introduced and used in France, Italy, New Zealand and parts of the United States.

Besides valuable research into alternative energy sources and routine use of alternative power, many devices such as solar power cells, methane digesters, recycling toilets, etc., are easily, commercially available. *Spectrum* (a special publication of the Alternative Sources of Energy group) lists hundreds of these. If you are interested in further exploration of the alternative-energy field, consult: (2), (3), (4), (5), (8), (9), (10), (13), (15), (16), (22), (23), (24), (32), (50), (51), (59), (63), (64), (65), (66), (71) and (78) in our bibliography. Most of these books have bibliographies of their own, some running to as many as three thousand titles.

This book is also *not* written for those who have opted out of society to live out some "noble savage" idyll in a rural commune in southern Colorado. "Dropping out" is a game for those with private means, regardless of their revolutionary rhetoric.

It is also *not* the intent of this book to merely evaluate existing products. Nader's Raiders and many consumer publications and organizations perform that function. We question whether the product should exist at all, or at least in its present form.

Nor have we written a fix-it book for garage or home-workshop mechanics. Their needs are catered to by *Popular Science* and *Popular Mechanics* and many other fine magazines. Titles dealing with this viewpoint will be found in the bibliography: (12), (14), (21) and (69).

We aren't saying that these groups are excluded from our book; rather we suggest that their more specialized needs can be better met by other publications. As we stated earlier, *How Things Don't Work* is a manual for coping with life in societies that are changing from an industrial to a post-industrial base. We wish you determination and luck.

Victor Papanek

N'Djamena (Chad), Guadalajara (Mexico), Ottawa (Canada), 1973–76.

James Hennessey

Los Angeles (California), Rochester (New York), 1973–76.

NOTE TO THE READER: We apologize for the fact that in evaluating products throughout the book, we seem to use a great many examples from the automotive field, as well as stereo equipment and cameras. This was done not because we are dazzled by their supposed splendor, but rather because the most ridiculous as well as the most advanced consumer innovations tend to show up first in these areas.

Acknowledgments

We want to thank Roger Dalton, Mitch Fry (from Arizona), Mark Hofton, Tim Lloyd, Mohammed Azali Bin Abdul Rahim (from Malaysia), Michael Morris, Alojzi Piatkowski, David Raffo and Mike Whalley, all post-graduate design students from Manchester Polytechnic; Tage Schmidt, a student from the Royal Academy of Architecture, Department of Industrial Design, in Copenhagen; Reinder van Tijen of the Royal Tropical Institute of Amsterdam. Photographs and drawings of their work appear in this book, as do many of their ideas. It was a remarkably fruitful two-way challenge to work with them.

We also want to thank Ira Velinsky, a student from the Rochester Institute of Technology, who helped with many of the drawings.

Most of all we wish to thank Sara Hennessey and Harlanne Papanek. Both were of constant help to us in discussing material and ideas. Their experiences, love and friendship provided an environment in which we could write with integrity.

How Things Don't Work

Throughout this book the sign ◖ at the left of a a paragraph indicates an alternative. In several cases this same sign accompanies a chapter heading, since we feel that some chapters constitute alternatives in their entirety.

1

How Your Bathtub Doesn't Work

It may be that what we call modern is nothing but what is not worthy of remaining to become old.

—Dante Alighieri

One of the proud citadels of the American housewife is the bathroom. Generally, there is agreement, even among thoughtful and emancipated women, that the bathroom *does* work and, except for shower enthusiasts, that the bathtub itself seems to be a faultless piece of equipment. That isn't so. The bathtub doesn't work. Neither does the shower. Nor does the toilet, or the washbasin. Contrary to popular myth, the room is unsanitary, dangerous, badly thought out, energy wasteful and a health hazard. It does glisten and sparkle (when it's brand-new) and it is certainly costly.

We will use bathtub and bathroom as a sample case history and employ alternatives to show how things might work better.

Since a tub must be big to feel comfortable to a reclining bather, it tends to be heavy. Often it is slippery (hence dangerous) and difficult to leave or enter. It is hard to clean if made of vitreous china; if made of fiber glass or other plastics, it needs special cleansers; and if made of enameled cast iron, it chips easily. Finally, it is permanently and wrongly placed. Its usual position under a window ensures a splendid collection of troubles. There are drafts, sending icy blasts down your back while you try to enjoy a nice warm bath. Then there is the question of visual privacy. One answer is windowshades, which, if the lighting is just right, can provide your neighbors with all the enjoyment of a classical Balinese shadow play. Another popular solution is frosted windowpanes, which lend an even sharper definition to the shadows projected for your neighbors' benefit. Some people prefer various curtains and drapes that usually combine the visual splendor of a Louis XIV damask with the rubbery chill of a used Glad Bag. If made of cloth throughout, there's nothing worse than soggy drapes hanging over your back.

Meanwhile, soot filters in through the window cracks, escalating the cleaning problems of the bathtub. And finally, under the continuous assault of water, the windowsill itself will rot under the chipping paint that slowly descends on tub and bather alike like melancholy confetti.

◑ A simple, do-it-yourself solution to the dangers posed by slipping and falling in the bathtub was developed in Guadalajara. Gerhard Kunze, a young Viennese designer, and his Mexican wife, Patricia, made a series of foam liners that attach partway down into the tub and over the sides and top. The foam pads are enclosed in waterproof plastic that is weighted to reduce buoyancy. This creates a padded tub, protecting small children, or anyone else, from injury. The pads are easily removed for cleaning, the foam itself is protected from water, the vinyl covering wipes clean easily.

◑ This raises the question of a manufactured alternative. There is no reason why an entire tub should not be mass-produced in self-skinning foam. It would be soft and pleasant to the touch, avert injuries, be nice and warm, lighter than most conventional tubs and hence easier to ship and

4

1 California "hot" tub
Photo Richard Fish. Reprinted from
Hot Tubs (Vintage Books).
Copyright © 1973 by the Capra Press.

less trouble to install, and no more difficult to clean than other plastic bathtubs. Finally, in case of damage, it would be easy to repair. The sensuous, yielding surface could come in a wide range of colors.

Then there are the tub controls. Hundreds are available, ranging from simple turning spigots to a matched pair of dolphins rampant in solid gold at $2850 the set. None of them work. Besides being difficult to understand, they constitute hazardous obstructions within the tub and also waste both water and energy. In a pamphlet prepared by Moen Products in Ohio, the understandably self-serving emphasis of their copywriter does not alter the facts:

Industry research has concluded that about 60% of domestic water which is wasted has been heated before pouring down the drain unused. Let's go back to our little exercise of water wasted by merely "fiddling" with two-handle faucets—the fiddle factor—7000 gallons each per year. 60% of that was hot—or

4,200 gallons—before it gurgled expensively down the drain. As a nice convenient yard-stick, and speaking in year-round national averages—most domestic water is heated about 100 degrees between the incoming water main temperature underground—and the hot water which comes out of your faucet. Typically the spread would be 50°F "In" and 150°F "Out." Whether using fuel oil or electricity or natural gas, domestic energy now costs about 15¢ to heat 100 gallons of water 100 degrees. By simple arithmetic, then, we've also "fiddled" away 42 (00) $\times$ 15¢ or $6.30 worth of energy per faucet—speaking conservatively. That's over 17 gallons of oil running down the drain along with the water—or a lot of watts through the meter, or gas through the pipe—for each of those energy wasters. The average home has five or six of them, endearing themselves to every Sheikh in the Middle East every time they pour another barrel of oil down the drain. On that shower example—36,500 gallons total water savings a year—about 22,000 gallons has been heated—or about $33 worth of energy also saved....[1]

Finally, tub controls tend to be ergonomically badly designed for the human hand, especially a wet soapy hand. Also, controls are usually located too far away to be reached comfortably.

The drain closure is equally difficult to reach. When mechanical, it is installed so that repair would baffle an experienced safecracker. The other solution tends to be a loose, clinking, rusting metal chain that gets forever tangled up with your big toe.

While the bathtub is slippery enough to have become one of the most common sites of accidents in the home, its relatively high sides and its general configuration make it nearly unusable for people who are elderly, partially handicapped, pregnant, obese or, for that matter, many smaller children. Handlebars that facilitate tub entry and leaving are seldom seen outside clinics and hospitals; instead we install extra "show-towel" racks.

The object in taking a bath is certainly not just to get clean. If it were, we could just put on a Mylar jumpsuit, zip it up to the neck, fill it with warm, sudsy water and roll around on the floor in it. Or else corporate sales technology would surely provide us within days or weeks with three gigantic aerosol cans containing: "Sudsy Water," "Magic-Rinse" and "Hot Air" under some label like "Navajo Desert Breeze."

Bernard Rudofsky has made a strong case in his books for the hedonistic pleasures of bathing and communal bathing.[2] Some of his ideas, after a prolonged cultural hegira via Big Sur, Hot Springs and Esalen (with dashes of the Finnish sauna and the Japanese ceremonial bath) have at long last emerged as the California "hot tub" culture. The California hot tub (see Figure 1), designed to be used by two or more people, is usually made of wood. (We used a cut-down Napa Valley wine barrel years ago.) Devotees of hot tubs assume that, as in the Japanese bath, no one would dream of entering a tub unless already clean, so the hot tub is dedicated to sensuous pleasure rather than scrubbing. Its natural setting is outdoors in the garden, or, if the climate doesn't permit it or neighbors are too close, in some indoor area—but *not* the bathroom. Sensory pleasure will obviously be negated if the tub is installed in a latrine, in accordance with our normal custom of putting toilet and bathtub in the same room. Special care must be given to ways of entering and leaving such tubs however.

For people less experimentally inclined, we have designed a nomadic bathtub made of a folding aluminum skeleton and a thick polyethylene liner (see Figure 2). It has some advantages over conventional bathtubs. You can build it to the size you desire—single, double or larger. (Caution: The weight of a full tub with two people

2 "Nomadic" folding tub

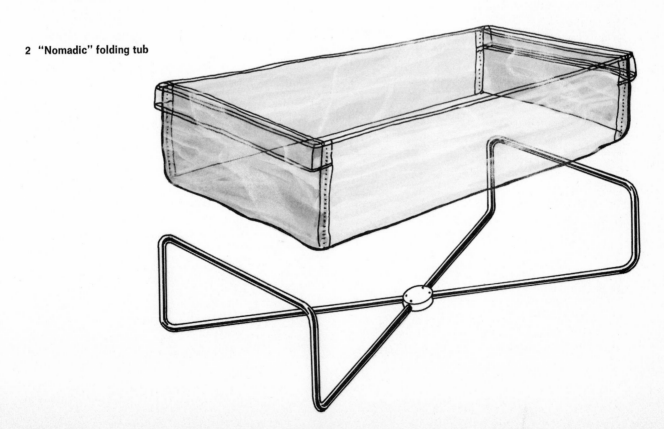

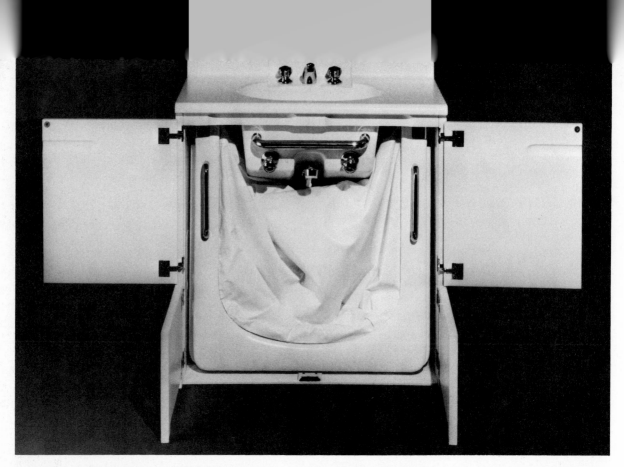

3 Babette Newburger's "soft" (folding) bathtub in its storage position under the sink. Patented by Babette Newburger

4 "Babette's Bath" as it emerges from under the sink. Patented by Babette Newburger

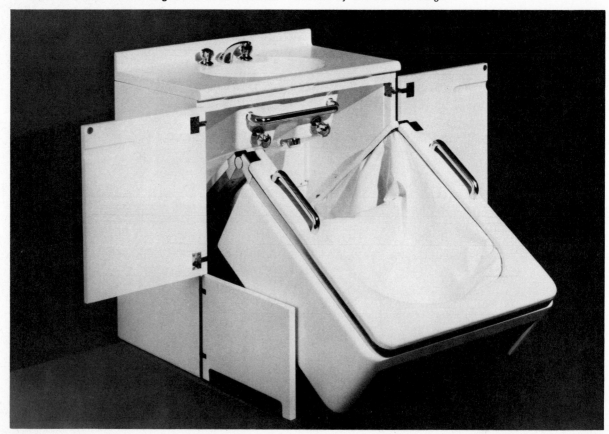

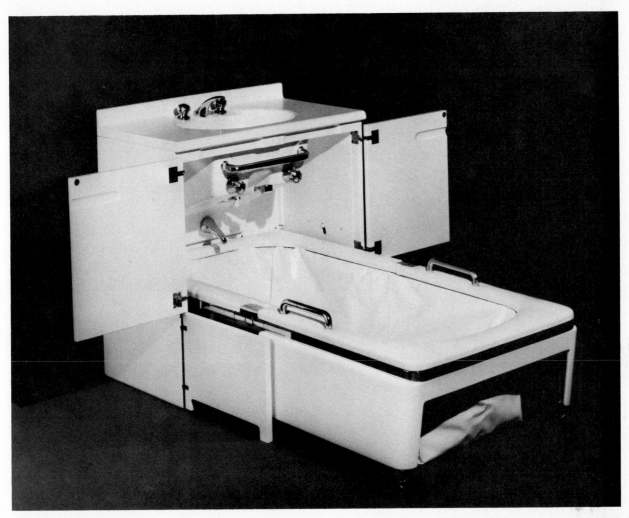

5 The soft plastic bathtub, fully extended—safe, comfortable, and easy to clean
Patented by Babette Newburger

in it will be about 2500 pounds; a single bath with one occupant under 1200 pounds, or slightly more than the weight of a comparable water bed.) Once built, it can be folded and moved. More important, it can be repositioned in the house or, local bylaws and neighbors permitting, moved outdoors. In seriously cramped quarters it might conceivably even be folded up and stored when not in use, thus providing more floor space.

Since 1970, Babette Newburger of New York has attempted to market her "soft" (actually. folding) bathtub, so far unsuccessfully. (See Figures 3, 4, and 5.) Her own hopes center around her tub, made of soft and inexpensive plastics, being used wherever portability is of importance: disaster areas, overseas construction sites, for handicapped people too weak to be brought to a permanently positioned tub. But she also hopes to place her tubs in places where none are at present available. She cites slum housing and low-income areas; we are also aware that in so socially responsive a city as Copenhagen, 21 percent of inner-city residents have no indoor bathrooms. Predictably, Mrs. Newburger found that plumbing manufacturers were totally uninterested in her design, since it would require them to switch gears and invest in new manufacturing equipment. (That's "free" enterprise, folks!)

In terms of entry and leaving, sunken tubs have an obvious advantage, especially if they are equipped with handrails. We recommend a sunken tub over a normally placed one, and add that if the tub can't be sunk, the floor can be built up, as shown in Figures 6 and 7. Figure 8 shows a typical Swedish tub which solves this problem in a still different way. It is stepped, to help small chil-

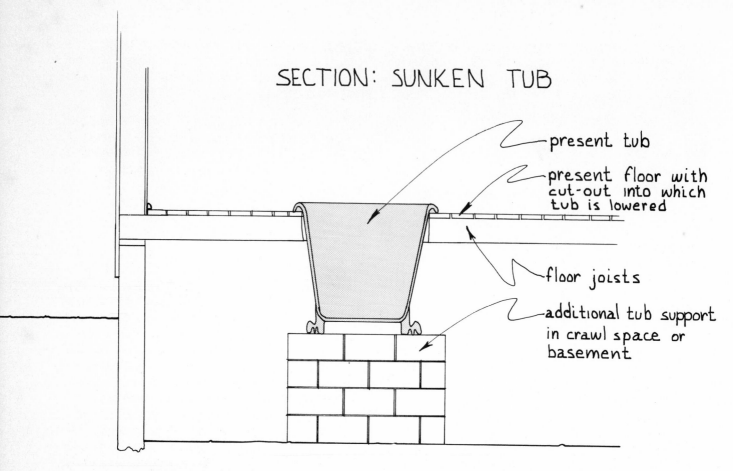

SECTION: SUNKEN TUB

present tub

present floor with cut-out into which tub is lowered

floor joists

additional tub support in crawl space or basement

6   Self-built sunken tub

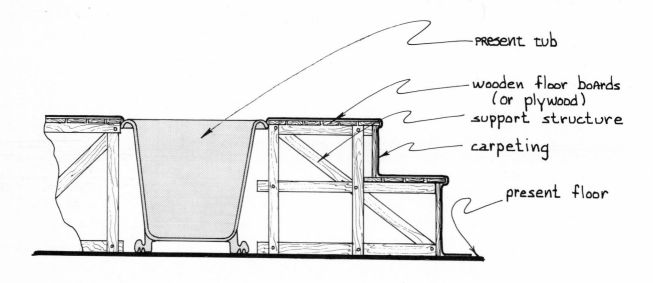

present tub

wooden floor boards (or plywood)

support structure

carpeting

present floor

SECTION: RAISED FLOOR

7   Self-built raised floor around tub

dren, elderly or obese people. This type of tub can be obtained directly from Sweden or through a few American distributors. It can also be built as a fiberglass lay-up. (Slightly larger ones already exist as personalized subtypes of the Hot Tubs in California.)

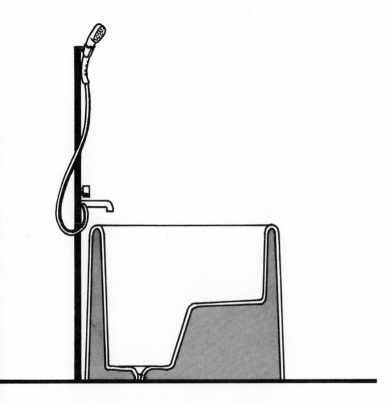

8 Typical Swedish tub with step and telephone-type shower

◉ Earlier we quoted extensively from a pamphlet by Moen that deals with single-handle controls. While we recommend such controls, this is still not good enough. The sudden cutting in or out of a major home appliance (dishwasher, washing machine, etc.) can drastically raise or lower the temperature of the running water. Photo labs (where constant temperature flow is important) solve this problem through a high-technology, over-designed sensing valve that continuously monitors the water temperature. We feel (as shown in Figure 9) that present technology enables us to construct a simple bimetallic sensor in a plastic housing that will maintain water temperature, once selected, for tubs, showers and basins.

◉ Drain closure can be handled in the simplest way with a four-inch-diameter flat rubber disk as made by Rubbermaid, or, as common in Europe, a hard rubber ball that forms perfect closure with the drainpipe.

Regarding the position of the tub under the window, we have no solution for repositioning it. The same wonderful open-mindedness that inspires architects to place revolving doors and steps in public spaces used by people in wheelchairs has caused them to put tubs where they're least useful. Aside from chatting with your architect or prospective builder, we have nothing to offer.

Showers are often part of the bathtub. In recent installations the upper part of the tub may be encumbered by a series of glass panels and glass sliding doors, to close off the entire unit when taking a shower. The fact that such a superstructure makes the bathroom appear even smaller than it is already is secondary to the real problems in the system. These tub enclosures are not only hard to install, hard to clean and extraordinarily expensive, they also limit access to the tub, permitting only a half opening at any given time. The interior space, usually ignored when taking a shower, becomes a truly awesome spectacle while taking a tub bath. One has a great sense of being enshrined.

Since steel parts rust, manufacturers have recently switched to aluminum frames for these enclosures. Helpfully the screws are still steel so that now only the screws rust. But unfortunately aluminum oxidizes, so some of the aluminum oxide dust settles in the tracks making them almost impossible to lubricate.

Most important, however, this whole idiot greenhouse affair is extremely dangerous. The shower door sliding out of the track and smashing your toes while the door itself shatters against your shoulder leaving you standing in a bathtub full of splinters may not happen frequently, but people do slip in bathtubs. They slip quite often. Hence it is unforgivable to provide a large sheet of glass into which wet, naked people are liable to crash.

Swinging rather than sliding shower doors merely add to the danger, constituting what insurance agents sardonically and precisely refer to as "attractive nuisances" for small children. Recently another type of shower-enclosure door has

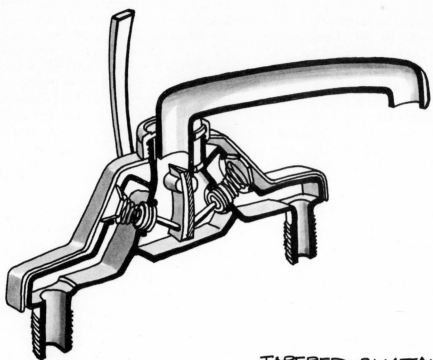

TAPERED BI-METALLIC WEDGE
IS DIRECTLY CONNECTED TO
HANDLE. AS HANDLE IS PUSHED,
WEDGE OPENS BOTH INNER VALVES.
THEREAFTER THE "SENSING"
BI-METALLIC WEDGE KEEPS THE
WATER TEMPERATURE CONSTANT.

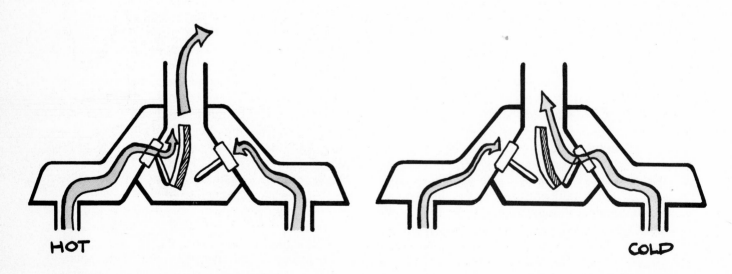

HOT

COLD

9 Suggested temperature-maintaining faucet, based on a bi-metallic principle

emerged which slides *up* instead of sideways, thus attractively combining the various hazards cited above with the concept of a do-it-yourself guillotine (for people whose bathrooms have sixteen-foot ceilings). Some shower enclosures now come in "shatterproof plastic." They don't shatter; they break and may expose sharp edges.

As for the showerheads, they are usually set into the wall so that they can dent a skull or poke out the eye of a tall adult, while being too high to be adjusted by children. If placed somewhat lower, one is forced to genuflect while washing one's hair. Showerheads are usually not adjustable except for one universal joint. In older models the choice is usually between dribble, drizzle and drip. But even newer ones get clogged with lime or other mineral deposits, turning the gentle spray into three or four needle picks that might be described as "aquapuncture." All in all the showerhead seems best fitted as a display rack for soggy pantyhose.

The plumbing for showers is buried in the walls, and it is there that leaks tend to occur. For a while these will slowly seep through the floor and into the rooms below. Once an angry and dripping tenant from below has alerted you to the fact that a problem exists, it can easily be fixed by gutting the wall, ripping out the pipe, making the repair, shoving the pipe back in and replastering the wall—until next time.

Shower controls are basically dangerous protuberances into the bathing-showering area. They are confusing because of the many different types available; one has to "learn" each new shower. They tend to be dinner-plate-size wheels, entirely too large for the human hand, or else a series of levers, handles, spigots and valves that get too hot to touch and too complex to understand.

Where the shower is in-tub or separate, we recommend shower curtains instead of glass and metal enclosures. Besides being easy to use, clean, maintain, and repair, they are also less expensive. These now are lined with soft "breathing" plastics on the inside, taking away the chilly sogginess alluded to earlier. While the glass/metal/plastic enclosures mentioned above are often "prettied up" with images that call up the Alhambra in Spain, we feel it is safer, cheaper and in many ways aesthetically more satisfying to surround a shower with just a plastic liner to which a good-looking fabric can then be added on the outside.

In Finland and the Scandinavian countries, and especially in Sweden, the entire bathroom is built around the shower (see Figure 10). Imagine then a room about five by six feet, the concrete floor of which is slightly concave leading to the drain in the center. The floor area is covered with a series of removable birch slats on frames. Walls are waterproof and washable, or, more rarely, half tiled. A "telephone" shower with controls in the handle hangs from the wall and can be used manually all over the body. Attached to a six-foot flexible hose, it can also be used to wash down the bathroom. Four or five hooks are provided for

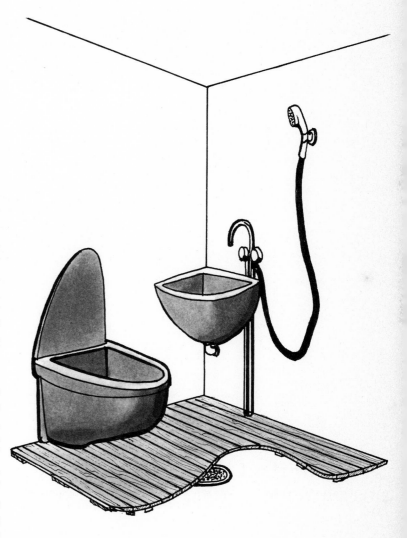

10 Typical Swedish bathroom with removable slatted wood floor, partially cut away

11 Telephone-type hand-held shower, attached to conventional U.S. plumbing. A simple, inexpensive attachment compared with . . .

the telephone shower at various wall heights. The washbowl, similar to those in North America, can however also be served by the same showerhead for shampooing, etc. A simple toilet bowl may complete this typically Swedish bathroom.

Telephone-type showerheads were recently introduced in this country and can be recommended highly. However, with typical North American exuberance for replacing what is simple and good with what is complex and costly, dozens of units with "variable needle sprays," "pulse-a-magic," vibrators and "rotary massage sprays" have been developed that are expensive to buy, quick to clog up, and of a size that would put a Kansas sunflower to shame. These combinations of garden watering can and electronic dildoe recall a comment made by M. M. Musselman in another connection: "When Detroit imitates, it is as though a hack artist copied the Mona Lisa, adding a Pepsodent smile, red hair and a rhinestone tiara."[3] And a "Naughty Nipples" peek-a-boo bra from Frederick's of Hollywood, we might add. However, simple ones are still available and we show one for $4.95 in Figure 11, contrasting it with a "de-luxe" (?) $69.50 model in Figure 12.

12 . . . "tools" such as these, which are beginning to be marketed now.

In terms of building, installation, maintenance and use-energy-cost, showers make more sense than bathtubs, both economically and ecologically.

Washbasins tend to be somewhat massive on the outside while interior space is oppressively small. Frequently basins come in sculptured shell-like shapes, and one sink specialist on 57th Street and Madison in New York City lovingly crafts sinks out of lapis lazuli, rhodocrusite, malachite or tiger-eye agate at between $7000 and $15,000 a throw.

The generous edge, holding soap debris and drippings, seems an unfair design trade-off for a larger basin, the joys of which are unknown to many. The average basin is so small that people are forced to shampoo and wash their hair kneeling next to the bathtub, in the shower or even in the kitchen sink. Basin heights are frequently arbitrary: sinks are too high for children, or far too low. Standardization, proceeding from ergonomic data, seems long overdue. The problem of sinks that are too high for children is elegantly solved by the Swedes, who have a fold-down step for children that is topped with a nonskid surface.

In Figure 13 we show drawings of a washbowl that is more adequate in size, deeper front-to-back and also uses a telephone-showerlike head. Bowls like these have become more easily available during the last two years.

The toilet in a 747 is in some ways better designed than that in the average home. The airplane's bowl has a rapidly sloping fall-away throat. Rinse water is dispersed in a circular spiral from above rather than below. No delicately tinted blue water fills the throat so that each act of defecation becomes a partial sitzbath.

Alexander Kira did a monumental study entitled *The Bathroom: Criteria for Design*.[4] The book goes into painstaking detail about sitting and standing urination procedures and, in its improved revised edition nine years later, prints all the photographic illustrations that were deemed culturally unacceptable a decade ago. While Kira's book is unusually thorough, it tells us more about bathrooms than we care to know. His design suggestions are all high-technology, mass-production answers, some of which could be obtained more simply.

Most toilets are too shallow in interior slope with a water-cleansing process that is quite ineffective as far as the interior of the bowl is concerned. Frequently the toilet is too high and designed sculpturally for aesthetics rather than function. Cleaning the exterior of a toilet bowl usually involves kneeling and embracing the whole zany contraption in order to give it a good rubdown.

The toilet tank tends to be outsize in order to maintain sufficient water pressure, and its interior workings are incredibly confusing nests of tubes, gaskets and wires that have failure and leakage engineered in. Although both seat and lid are designed to be sat on, over the last ten years the plastics used for this have become thinner, more

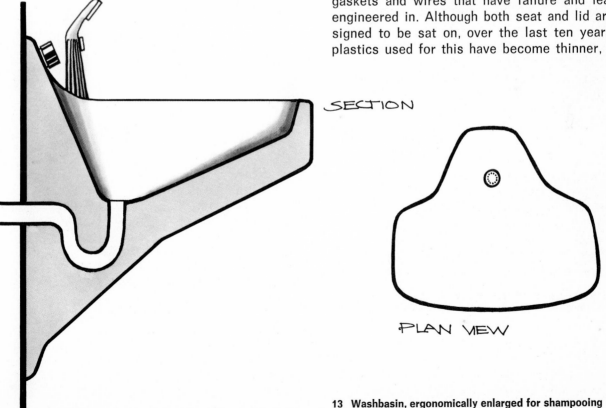

SECTION

PLAN VIEW

13 Washbasin, ergonomically enlarged for shampooing

14 Typical European toilet, showing
rapid fall-away throat

flexible, lighter in weight and readier to dislocate
or break. Hence the amazing sales of toilet-seat
covers ranging from simple shag rugs to needle-
work bicentennial comforters to imported monkey
fur and chinchilla.

The development of a selective flush toilet
with different volumes of water for urination
and defecation is comparatively simple and was
suggested by the author in 1969.[5] Present-day German and Scandinavian toilets, usually with stain-
less-steel interiors, possess both a rapid fall-away
throat and strong water-cleansing action from
above (see Figure 14).

Just by wall-mounting the tank high on the
wall, greater water pressure with less water
is achieved through gravity. This type of system
(Figure 15) has existed in the United States since
the turn of the century and has fallen into disuse
only since plumbing-fixture manufacturers began
hiring sculptors with degrees in industrial design
to "improve the image." Since toilet seats can be
bitterly cold, materials kinder to the human skin
might be reintroduced, such as unlaminated wood.
Presently, when considering the bathroom as a
whole, we will be showing combination sink and
toilet systems as well as combination sink/toilet/
shower systems.

While the toilet bowl itself remains inefficient
in the ways we have explained, marketing people

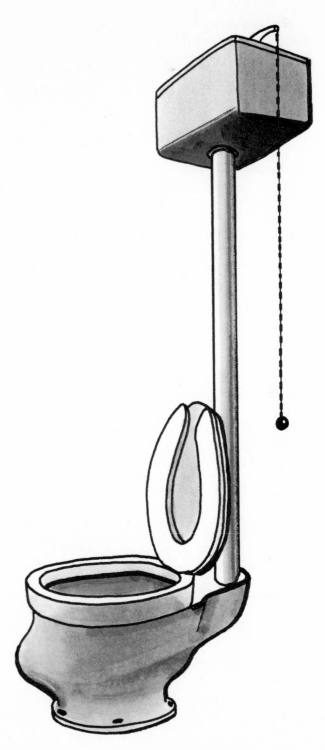

15 Old-fashioned U.S. toilet with high, wall-mounted tank

and the innovation brigade are forever busy dreaming it into a greater state of preposterousness. Within the last five years national magazines carried advertisements extolling the virtue of a "paperless, electric toilet seat." The device, continuously circulating warm water to keep the seat warm, would use a warm water stream and hot air. After completing the act of defecation, the device would be activated. A stream of water, presumably at room temperature or warmer, would "rinse," then a jet of warm, forced air would complete the drying process. If memory serves, it was to sell for around $250.00.

The mirror above the sink is eminently practical. However, the medicine cabinet behind it is not. Because of architectural "aesthetic" considerations that strive for flush mounting, medicine cabinets tend to be unusually shallow and also narrow to fit between studs. With our national fetish for patent medicines, nostrums, salves, unguents and beauty preparations, we tend to stuff the cabinets till they burst. Opening the door then causes a surrealistic clutter of birth control pills, Gelusil, Helena Rubinstein's "Command Performance," toenail clippers, eye lashes and dental floss to cascade into the sink, breaking bottles and jars.

This hazard, however, is not nearly as great as the fact that medicine cabinets are not childproof. We feel strongly that for safety and health reasons medicines and other drugs must be stored somewhere else. With children in the house, or the likelihood of children visiting, such materials should be kept in a childproof and adequately spacious cabinet that can be locked or, preferably, has some sort of device that foils the exploring hands of a small child.

The compulsory introduction of "childproof" containers (Figure 16) for all prescription drugs in the United States, West Germany, Canada and a few other countries has not entirely solved the problem. One of our graduate students designed the first prototypal "Pill-safe" in 1965.[6] Since then lid-controlled, less expensive pill jars have been made, but now children still manage to poison themselves by drinking bleach, detergents and other household chemicals, or by chewing on fabric-softening rags. The obvious answer lies in a cabinet that stores all such materials and is lockable.

However, a chest locked with a key or combination lock isn't the right answer either. Often an elderly family member may need a heart-ailment remedy quickly, without fiddling with locks and keys. Or a severely arthritic grandfather, say, may be unable to operate a key. To solve the problem, our students in Denmark designed chests that took advantage of children's inability to spread their tiny hands across a given space, yet permitted adults, even severely crippled ones, instant access. A news conference was held in Copenhagen and a half-hour TV show interviewed some of the students. As a direct consequence of the newspaper and TV coverage, passage of a new law seems probable in Denmark, making such cabinets (with ergonomically designed latches that are childsafe) compulsory in all apartments and houses rented or sold.

We have been able to push this concept somewhat further, so that the government of New Zealand is also considering steps to make such a safety cupboard compulsory. The Standards Asso-

16 Prescription-drug lids that are "childproof"

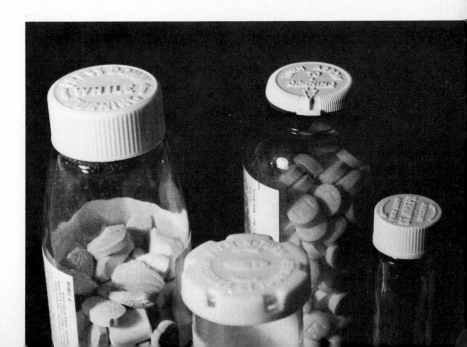

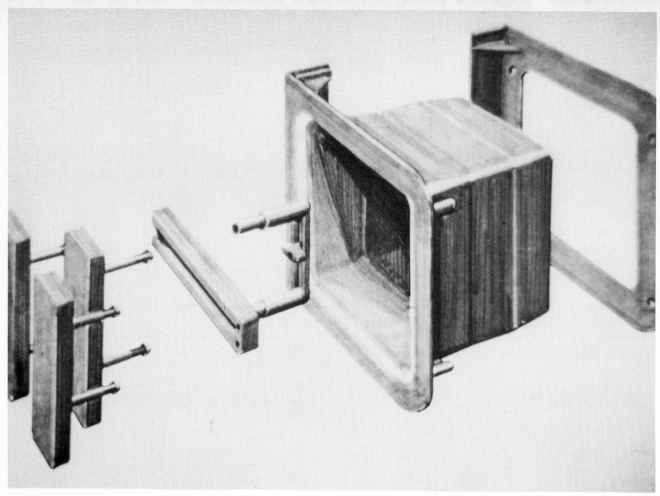

17 Sketch for childproof, install-it-yourself cabinet lock, ergonomically designed by Tim Lloyd, a postgraduate student at Manchester Polytechnic

18 The childproof lock before installation

19 The childproof lock installed

ciation has prepared a draft amendment to its model-building bylaw which, if tested and approved, will require a cupboard that must be installed in all new homes and flats, and cannot be opened by children. The same cupboard must be "lockable" without a key, thus giving access to adults who are severely arthritic or in other ways need fast access without fumbling for keys.[7]

In Britain, Tim Lloyd, another postgraduate student, designed a do-it-yourself cabinet lock (Figures 17, 18 and 19) that is easily installed and will defeat any child's attempt to open it. This lock will convert any existing chest, cabinet or storage unit into a "safe" area in a few minutes. In an earlier book, we have shown even simpler do-it-yourself conversions.[8]

The bidet, long kept out of American bathrooms as a result of our puritanical heritage (in fact, considered so "naughty" or "decadent" a concept that it appears neither in the Random House American College Dictionary [1963] nor Webster's 7th New Collegiate Dictionary [1966]), is finally be-

ginning to be acknowledged on this side of the Atlantic as a normal sanitary device. We may expect product designers to give us the biggest, jazziest bidet in the world before the end of the decade.

In some parts of this country where the climate is usually mild, bathrooms have electric heaters or people drag ancillary electric heaters into the room. Since water and electricity don't mix too well, we suggest the following alternatives:

Electric heaters should be permanently installed and protected, or other heating systems should be used (hot water radiators and baseboard units, forced air or infrared lighting). Since there is also no adequate seating in most bathrooms, we have sketched a combination birch slat seat and heater enclosure from a Swedish bathroom (see Figure 20). It has the additional advantage that the warm air rises and warms the bench. The unit is childproof.

Bathroom scales don't work. Being spring-balance scales, they give inaccurate weight com-

20 Swedish bathroom, showing slatted bench with heater underneath

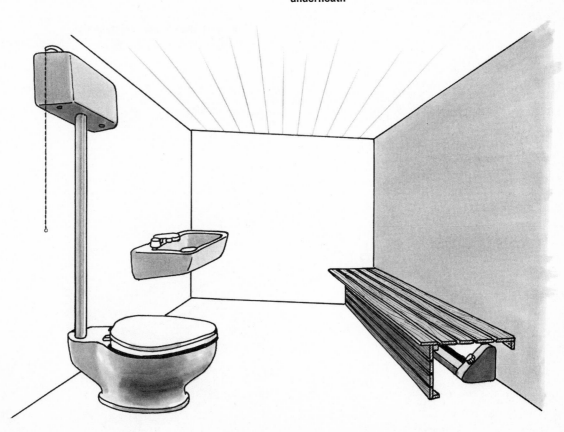

21 Toilet and tank of the Hübner and Huster modular bathroom system (exploded view)
Courtesy *"md"* Magazine, Germany

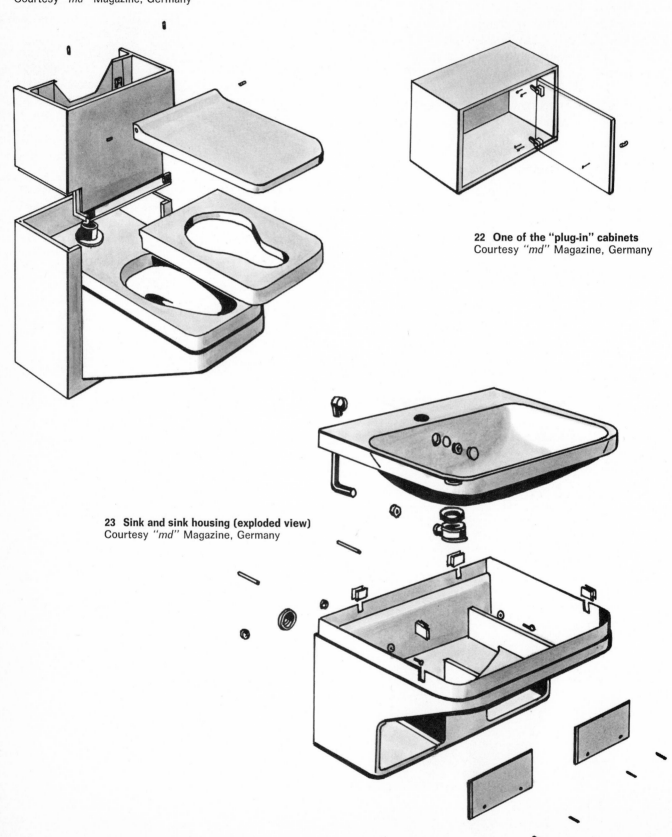

22 One of the "plug-in" cabinets
Courtesy *"md"* Magazine, Germany

23 Sink and sink housing (exploded view)
Courtesy *"md"* Magazine, Germany

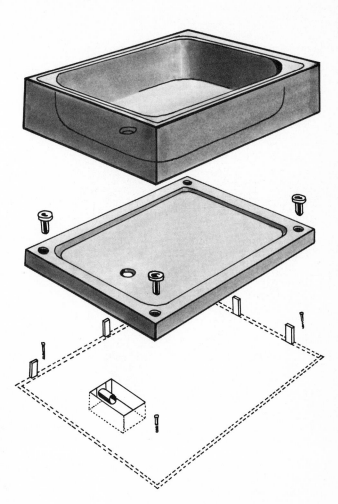

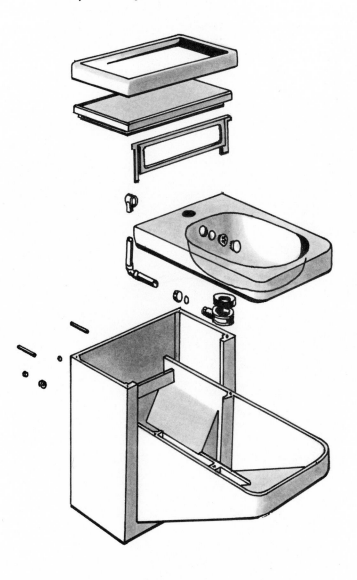

24 Bidet (exploded view)
Courtesy *"md"* Magazine, Germany

25 Shower basin (exploded view)
Courtesy *"md"* Magazine, Germany

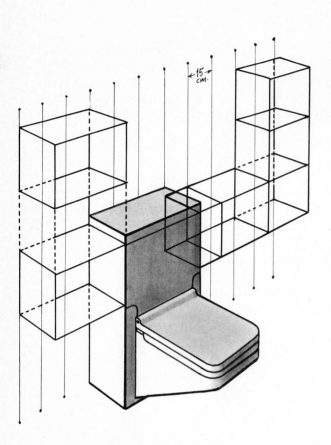

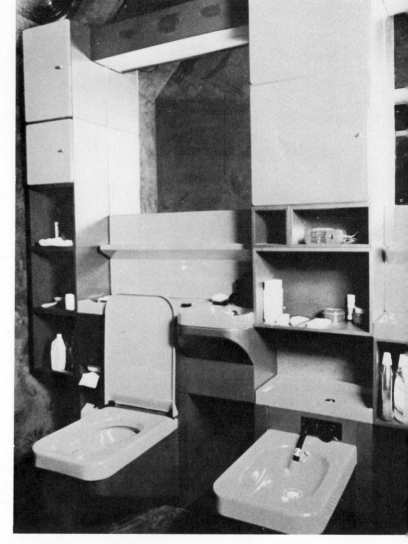

26 Schematic drawing, showing toilet placed with some storage elements
Courtesy *"md"* Magazine, Germany

27 The system installed
Courtesy *"md"* Magazine, Germany

pared to the beam-balance scale used by doctors. The springs themselves wear down unevenly, and in most scales they are not sufficiently protected from water spilled in the bathroom. Once rust gets started, their marginal accuracy deteriorates further. Most absurdly, the weight indicator is difficult to see by pregnant women, grossly obese people, the elderly and others with bad eyesight (for example, obese diabetics): precisely those groups that are supposed to keep a constant check on their weight.

There is great need for a low-cost, accurate balance-beam scale that knocks down. This could be simply developed by designer and manu-

facturer, and might even be available in kit form.

The bathroom doesn't have to be approached on a piecemeal basis, however. It can also be seen as a total system. Designers and architects have been exploring just such a total design concept for a very long time. Buckminster Fuller developed a two-piece molded "Central Utility Core" bathroom for many of his early synergetic and geodesic designs as well as for the "Dimaxion" house of 1946. Mosche Safdie developed a molded bathroom for the "Habitat" cluster of buildings in Montreal. Unlike other visionary total-bathroom approaches, his were actually produced by the

Crane Company for Canada. However, only 540 units were built and then production shut down. The one-piece bathroom has floundered because of crafts union agitation against it, as well as the fact that industry considers it more profitable to sell individual appliances. This seems a pity, for with the reclamation and upgrading of old houses and inner-city areas, one-piece, plug-in bathrooms could provide a sanitary and inexpensive answer to renovation problems.

◐ In Germany, where such considerations have less influence on the market, the factories of Röhm Chemicals of Darstadt have recently introduced a plug-together "building-block system" bathroom. Designed by Hübner and Huster, it is called "Plexmobil" and comes in orange, yellow, beige, green and blue. Its many modular components, all based on a 15cm increment grid, may be wall-mounted, built in, or floor-based. Devised as a system for fitted furnishings of almost any kind in the way of toilets and bathrooms (even changeable after the completion stage, if you wish), it comprises twenty-five individual types of components and accessories such as bathtub, shower basin, washbasin, toilet bowl, urinal and bidet as well as a variety of cabinets, shelves, mirrors, lights and back panels. (See Figures 21 through 27.) The simplicity and logic of Plexmobil's plumbing connections make the whole system ideal for the do-it-yourselfer. Because of its design logic we have fully illustrated it.

◐ On a much simpler level we show a free-standing unit combining washbasin and toilet in Figures 28 and 29. Used wash water, stored in a holding basin just below the sink, is used to flush the toilet, as it provides adequate amounts of water and pressure from gravity. Not only can water from the washbasin be reused, but solid and liquid waste can also be recycled.

◐ While alternative-culture magazines feature conversion plants for making methane gas from chicken droppings, a Norwegian company has developed a toilet costing about $400 that includes an attached freezer that solidifies the waste to prevent odors and bacterial action. The toilet uses electricity but not water or chemicals. The waste matter is stored in a biodegradable plastic bag that can eventually be composted.

◐ Since the mid-fifties a Japanese company, Toto Ltd., has combined a washbasin with the toilet, similar to our design suggestion above. The result is a free-standing unit that uses water from the sink mounted on the top of the toilet tank for flushing. Water saving from this combination is 25 percent, in addition to savings in cost and space.

◐ We suggest that the same effect can be achieved with conventional fixtures. The Minimum Cost Housing Group at McGill University has combined shower, toilet, washbasin and a European-type line heater in another free-standing unit (see Figure 30).[9]

◐ The Swedish Clivus Multrum is a toilet that successfully composts all human organic wastes without water, chemicals or electricity. More than 2000 of these devices are in use in Finland and the Scandinavian countries and are operated with the official approval of the Swedish Ministry of Health. Clivus owners in Sweden even receive a tax rebate, as they reduce the cost of sewage and garbage collection.

The Clivus itself is a fiber-glass container nine feet long, three feet wide and about five feet high. It is divided into three compartments: the top one for human waste, the center one for vegetable

28 Recycling washbasin and toilet unit

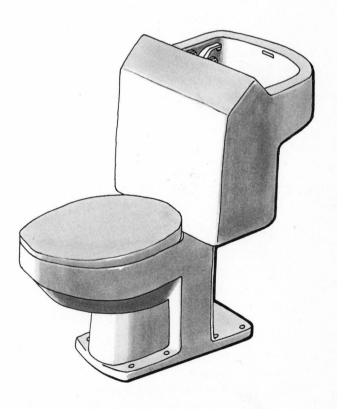

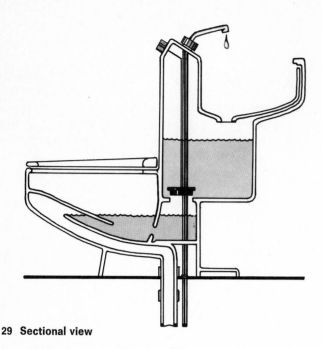

29 Sectional view

just a few of the things we haven't even touched. It's true that over the last few years many attempts have been made to "upgrade" the bathroom into a family area for pumping the old Exercycle or mastering the rowing machine. One can now enjoy a luxurious hydromassage through the use of whirlpool agitators with scented "salt" water in newer, squarish tubs. But these euphoric creations are devised for a well-heeled elite. If examined closely there is the same absence of trying to solve the basic problems.

We hope that you will arrive at your own solutions from a real appraisal of your needs, not the pages of some glossy shelter magazine.

scraps and other organic refuse, and the bottom one for holding the finished compost. A screw transport moves wastes so that the toilet (even several toilets) and the compost chamber can be mounted on the same level. The heat created by decomposing organic matter rises through a vent pipe, thereby creating a strong downdraft both at the toilet and at the garbage receptacle, making it odorless.

In order to get composting started, the bottom of the container is lined with organic material such as garden soil, peat or grass clippings. After the initial loading this process continues indefinitely, producing several buckets of rich humus per person, per year. The newly formed rich soil in the bottom chamber can be removed about once a year after a start-up period of about two years. Thoroughgoing tests by the Swedish Health Department have determined that all harmful bacteria, viruses or parasites are killed by a year of heat and bacterial action produced by the composting process. The unit is now also produced in the United States and is available from Clivus Multrum USA, Inc., of Cambridge, Massachusetts.

Even this lengthy discussion of bathtub and bathroom leaves much else to be considered. Chill floors and inefficient lighting, a shortage of counter space and storage areas, noise and odor factors are

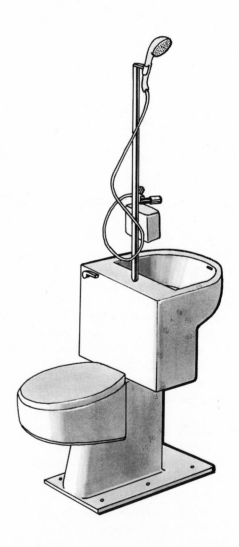

30 Free-standing recycling unit combining washbowl, toilet, telephone-type hand-held shower and line heater

2

Share Your Lawn Mower, Lady?

. . . A sense of community probably derived here from the stable-state notion that the world is a decent and satisfactory place which, if we share and cooperate, will sustain us despite some difficulties.

—Ernest Callenbach, *Ecotopia*

Even at the height of the growing season, your lawn mower is not used 156 hours each week. Chances are your vacuum cleaner gets used five hours a week and stands idle for 163 hours. Your freezer is probably 40 percent empty at any given time.

All of this must sound familiar, because that's the way things are in middle-class North America. When we moved to Ottawa, Canada, a year ago, we found ourselves on a typical middle-class, middle-income street. It was easy to do a quick survey: twenty-six houses with twenty-six families, but with twenty-eight lawn mowers (including two "riding-tractor" types and one manual one) and thirty-two vacuum cleaners (including two basement "shop-vacs" and several battery-driven miniature vacuum cleaners for car interiors).

Assuming four hours of use per family for a lawn mower per week during the summer, and eighty-four summer daylight cutting hours a week, one lawn mower could easily serve as many as twenty-one families. Even conceding that the lawn mower cannot be continuously used, that people are at work, that there will be a number of rainy days, etc., one lawn mower can easily be shared by five to eight households.

Since neither lawn mowers nor vacuum cleaners nor any of the many other things we buy and use sporadically are cheap, one of the questions that might well be asked is: Why not share your vacuum cleaner? The reason we're unwilling to do so is simple: We are afraid that it will be returned in a damaged condition. And we are probably quite right. It's not that our neighbors are clumsy or careless; it's just that none of these devices were designed to be shared with other people or to take abuse and punishment. The crux of the matter, though, is that most of us are still not very straight about the things we need to own and the things we only need to use.

In this context we could start with questioning the ownership of a lawn itself and what a front lawn requires. The typical front lawn, with all the attention lavished upon it in terms of automatic sprinkler systems, mowing, cutting and trimming, was used by Thorstein Veblen a mere sixty years ago as a prime example of conspicuous consumption and status seeking.[1] The lawn, of course, exists all over the world, except that in many places —Mexico, Spain, Italy, France—it is usually walled in and becomes an extension of the living room.[2]

The California life-style has at least restored the use of the backyard to us. As a rear patio it is used for sunbathing, barbecuing, entertaining and useful gardening such as growing vegetables. But the front lawn still lies there serene, beautifully manicured, useless, well cared for and expensive. In new subdivisions front lawns are frequently *installed* (a precise descriptive term in this case) by the speculative builder. A common method is to excavate the ground to a depth of about one and a half feet and then line the ground with polyethylene sheeting, divorcing the eventual lawn completely and permanently from the biosphere. After the polyethylene lining comes a shallow bed of gravel, across which the sprinkler pipes are laid, then a few inches of humus, some chemical fertilizer and grass seed. At the last moment the entire confection can be covered with greenhouse-grown Kentucky bluegrass turf.

Tract or development communities often have regulations that restrict the kind of ground cover that can be grown. And it may be hard to believe that only ten years ago a couple in San Mateo Township in California was subjected to a poison-pen letter-writing campaign, had their car tires slashed and their children ridiculed in school—all because they had decided to grow ivy as a ground cover, instead of grass.

Certainly our first advice would be to try to take possession of your own front lawn again by enclosing it and, local climate permitting, restoring it to its logical role as an outdoor living room and garden. Other possibilities are alternative ground covers: ivy, clover, myrtle, ice plants, succulents and many others. But if local rules and regulations still make the front lawn *de facto* "communal property," then it may be necessary to investigate communal action for dealing with the problem.

Forgetting for the moment about computer-steered watering and sprinkling systems (with humidity sensors as overrides to the sensing devices), the basic tool is still the lawn mower. Lawn mowers as a class are expensive and quite unsafe. Lawn-mower-caused injuries commonly occur when children manage to get their feet or hands under the cutting blades, or when rocks are kicked back and lodged in the person using the machine. Other accidents are caused when people have a trial run of their mower in a closed garage and carbon-monoxide themselves to that Great Front Lawn in the Sky. And then there's the individual who fills the gas tank with a lit cigarette negligently drooping from his mouth. Electric lawn mowers can create different problems. By running over the power cord on a wet lawn it is certainly possible to electrocute yourself, or at least incur the expense of a new power cord.

Lawn mowers in general are shoddily built. Engines are hard to start and require constant maintenance (at a higher mechanical maintenance level than an automobile), and they run poorly, if at all. Valves burn easily. Pull starters tend to break, sending the owner sprawling into the freshly ferti-

lized lawn. Blades nick and dull easily, which puts the mower off balance and causes it to vibrate. This in turn leads to pieces of the housing working themselves off and sometimes getting caught in the blades. Lawn-mower controls tend to look like bicycle brakes and gears, but whereas bicycle parts are weatherproof, most lawn-mower controls are not. Rod and wire components will rust, bend, buckle or shear. Wheels can sometimes spin off, due to vibration. Finally, a badly tuned engine can fill your trouser cuff full of sparks from the muffler, or start a minor grass fire.

Lawn mowers also pollute. While the industry is making honest efforts to reduce noise emission and "blimp" motors, the noise level of four or five mowers shrieking simultaneously on a Saturday morning is still unacceptably high. Of course, anything that emits exhaust will cause atmospheric pollution, and anything operating on electric current will use up energy.

Since the sun grows the grass, why not let the sun mow it? Considering the short use time of lawn mowers (see above), it is well within the state of present technology to develop a solar-charged battery-powered lawn mower. During the summer, a week's sunshine is sufficient to illuminate a low-powered bank of solar cells which in turn trickle-charge the battery in the mower (see Figure 31). The battery/solar-cell unit would be removable for constant exposure to the sun by storing, say, on a windowsill. A week's sunlight would easily provide enough power for a three-hour mowing job, but not much more than that. This may have the beneficial side effect of eliminating lawn freaks who constantly overmow their lawns.

Riding lawn mowers are the final absurdity. Though a man trying to cut his lawn can avoid exercise, pollute the area and feel like John Wayne, all at the same time, riding mowers have all the problems of other mowers on a more complex scale. (Recently *Popular Science* magazine suggested putting headlights on a lawn mower so that people could cut their lawns at night—a thought [considering how noisy they are] that would give pleasure to the late Marquis de Sade.)

Staying within the narrow confines of the lawn mower as a personal tool, some alternatives can be suggested:

Cordless electric mowers are being introduced as this is being written. The first of these, Toro's "Carefree Electric," is a battery-

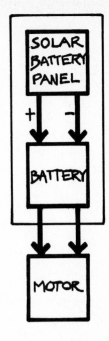

SOLAR CELL PANEL TRICKLE-CHARGES THE BATTERY DURING THE 1-2 WEEK INTERVAL BETWEEN MOWING.
REMOVABLE BATTERY/ SOLAR CELL PACKAGE CAN BE PLACED IN A WELL-LIT SPOT WHILE MOWER IS STORED.

31 Suggestion for a solar-recharged, battery-powered lawn mower

powered rotary mower which is unusually quiet, compact and lightweight. The machine was developed for small lawns and will cut for 45 minutes without a recharge (7000 square feet of lawn). The battery will take two hundred recharges before replacement becomes necessary.[3] Others are becoming available.

Deescalating the whole question from its present high-technology level: What's wrong with old-fashioned push-type lawn mowers? Recently we watched a gentleman ride around for an hour and a half on his "Wagonmaster Special" and then carry an Exercycle out on the front lawn and begin to pump it vigorously! The advantages of the muscle-powered push-type lawn mower are that it provides good exercise; it also cuts the lawn more slowly, therefore more carefully; and it is a simpler tool, hence much less can go wrong with it, and what does go wrong can be fixed easily. There are push-type lawn mowers that have been in continuous use for over sixty years and are still working well.

Finally Frank Rowland Whitt and David Gordon Wilson cite a pedal-driven lawn mower designed by Michael Shakespear at M.I.T., which we show in Figures 32 through 34. Their comments bear repeating:

The rationale behind the design of the grass-mower was that the leg muscles would be used more efficiently in pedaling rather than in pushing a regular lawn mower, and the back and arm muscles would be relieved; that continuous mowing would be more efficient than the frequently used to-and-fro motion of push mowers; that a multi-ratio gear would enable individuals to choose whatever power-output rate suited them and would enable moderate slopes to be more easily handled; that shortages of gasoline and anti-noise restrictions might limit the use of power-mowers; and that riding a pedal-mower might be fun as well as good exercise.

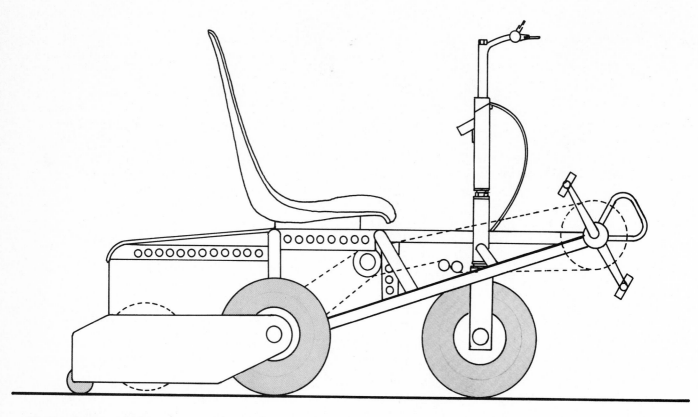

32 Muscle-powered lawn mower, designed by Michael Shakespear, a graduate student at MIT (side view)

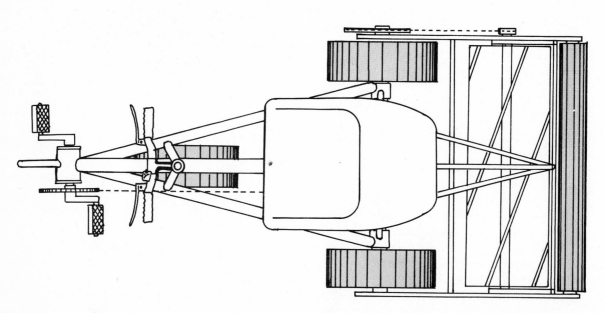

33 Plan view

The original model [shown in our sketch] has a three-speed Sturmey Archer hub-gear, a brake, and differential incorporated into the transmission. The reel-type cutter is driven directly from the input to the differential drive to the rear wheels. Pulling the left-hand handlebar lever releases a catch and enables the cutter assembly to be raised by pulling the handlebars back to a rear position and so permits easy maneuvering. The prototype, constructed largely of scrap materials and components, was very heavy *but still gave easy cutting. A light-weight model might show real advantages.*[4] [Italics supplied.]

The options that we have so far discussed (including a ground cover other than grass, or walling in the garden), are straight-line alternatives to things as they are now. What has not been sufficiently explored is the basic concept of *sharing.*

Few tools in our society are designed for communal (or shared) ownership. If they were designed for sharing, rather than for individual use, we believe they would change structurally, mechanically and in material composition. When reel-type push mowers were still in use, for example, they were more frequently borrowed or shared than today's electric or gasoline-driven mowers for the simple reason that they were robust, uncomplicated and difficult to break.

☯ In Figure 35 we show an electric lawn mower designed to be shared by six to eight families and developed with the help of Mike Whalley (a postgraduate design student at Manchester Polytechnic, in Great Britain). It differs in subtle but important ways from lawn mowers normally available for individual ownership:

1 The machine is made of heavier-gauge metal and more sturdily built, to take greater punishment by different people who may handle it.

2 The housing is surrounded by three heavy bumpers to protect it against obstacles.

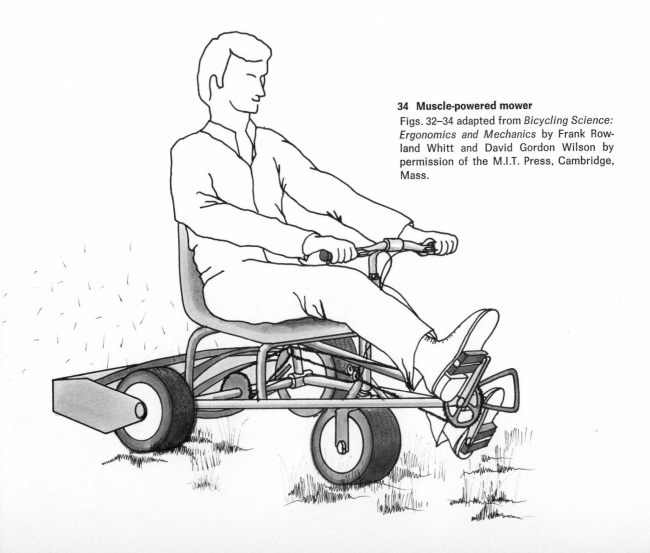

34 Muscle-powered mower

Figs. 32–34 adapted from *Bicycling Science: Ergonomics and Mechanics* by Frank Rowland Whitt and David Gordon Wilson by permission of the M.I.T. Press, Cambridge, Mass.

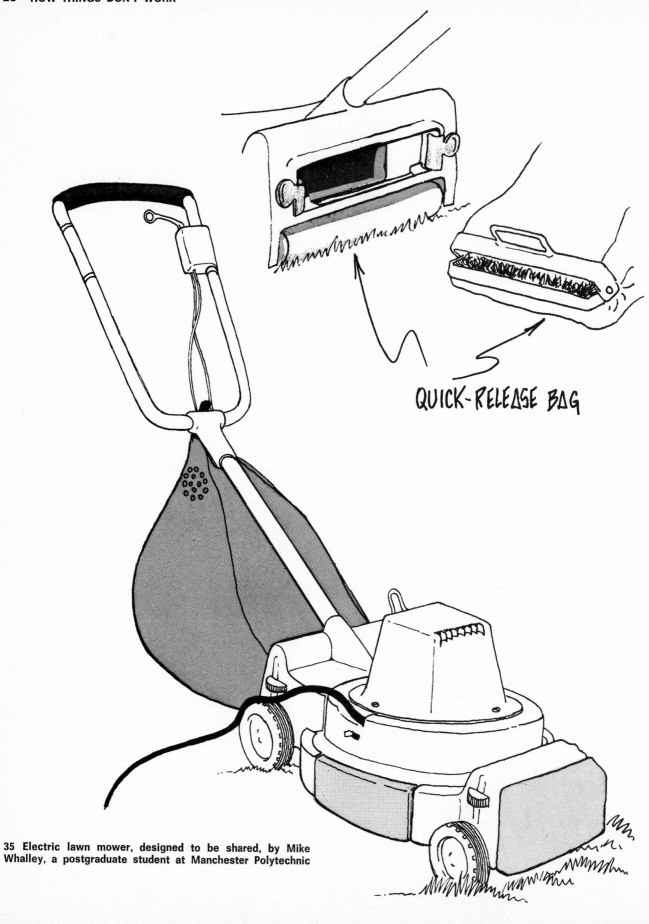

QUICK-RELEASE BAG

35 Electric lawn mower, designed to be shared, by Mike Whalley, a postgraduate student at Manchester Polytechnic

3 The grass bag is twice the "normal" size, easily detachable and made of biodegradable plastic so that it can go directly (with contents) to the communal compost area. In addition, as a safety feature, the bag has air holes to prevent children who might play with these bags from suffocating.

4 Instead of a fixed metal blade, the mower uses a heavy-gauge nylon fishing line for cutting. The line is stored in a small drum at the center of the motor shaft and a protruding length of it is "whipped" around to effect the task of mowing. As the line becomes shorter from shearing at the tip, more is simply pulled out of the drum. The great advantage of this system is that while the line is strong enough to easily cut grass, it will not cause serious injury should someone inadvertently put a foot or hand in it. Secondly this method eliminates the "off-balance" problems of conventional machines and the vibration that goes with it. Similar systems are just now being introduced by manufacturers on a small scale.

5 Cutting height is adjusted by raising or lowering the container drum of the nylon filament line. This permits the use of permanently installed wheels with full axles.

6 Hand-control clips are on either side for left- or right-handed people; control cables are heavily insulated and waterproof.

7 The entire motor assembly can be lifted out for immediate replacement in case of malfunction, and a new one "plugged in" while the malfunctioning, unplugged component is repaired.

8 The entire unit is double-insulated to prevent electric shock.

9 The power cord is retractable.

10 For safety reasons, the on-off switch is a *momentary* switch: It is in the "on" position only while the grip is held.

Such a lawn mower, designed for sharing, could serve present needs in a variety of different ways:

◖ The simplest and most direct way would be for six or eight families that live in immediate proximity to own it jointly and use it cooperatively.

36 Vacuum cleaner, designed to be shared, by Mohammed Azali Bin Abdul Rahim, a postgraduate student at Manchester Polytechnic
Photo Kokon Chung. Courtesy *Design* Magazine, England

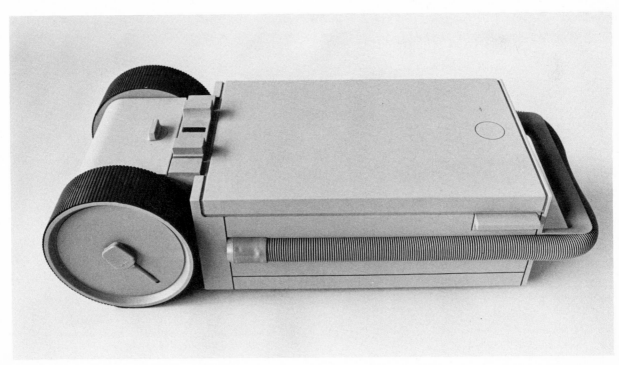

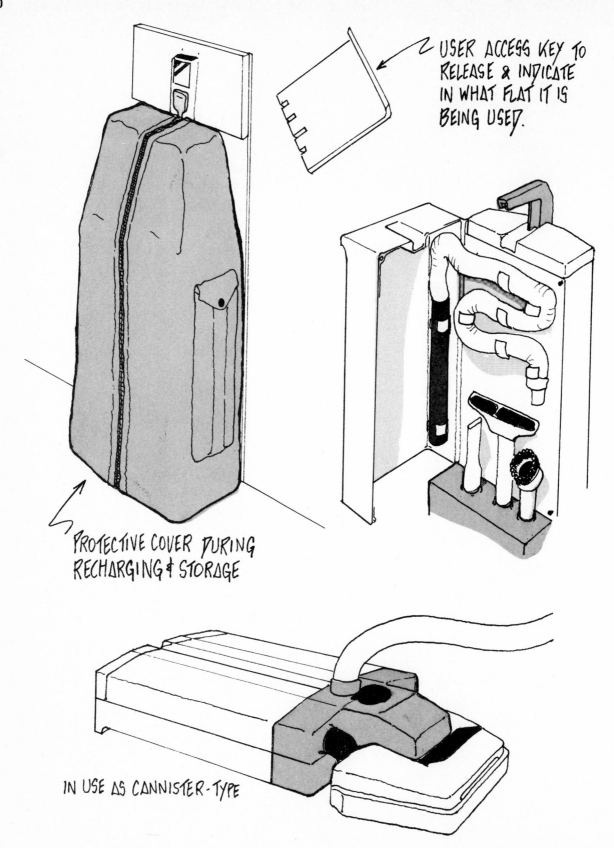

USER ACCESS KEY TO RELEASE & INDICATE IN WHAT FLAT IT IS BEING USED.

PROTECTIVE COVER DURING RECHARGING & STORAGE

IN USE AS CANNISTER-TYPE

Rechargeable battery-operated vacuum cleaner, designed to be shared, by Mike Whalley. 37.

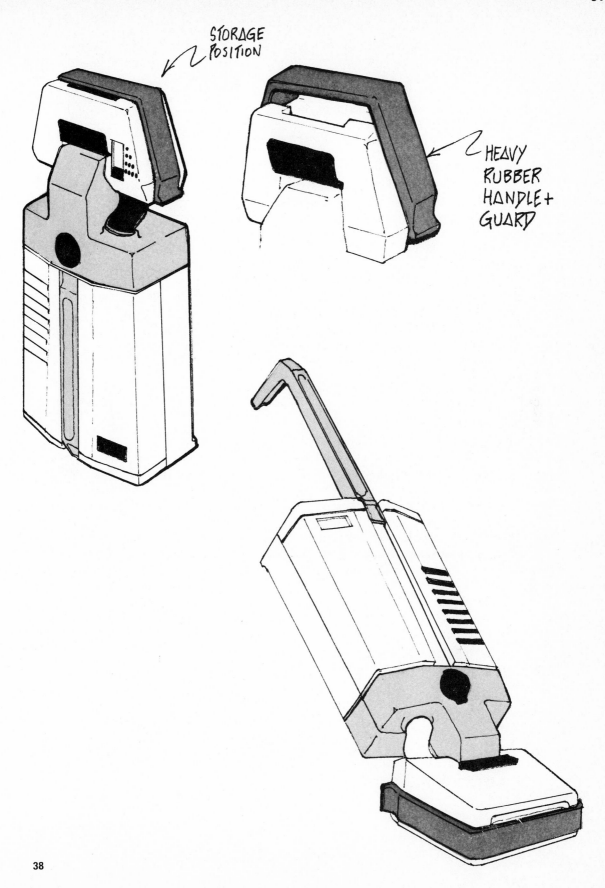

STORAGE
POSITION

HEAVY
RUBBER
HANDLE +
GUARD

It might be rented from a central facility within a small community that takes care of maintenance, repair and storage.

It could be used by a community-service group which, in addition to maintenance, repair and storage, also does the grass cutting.

At the beginning of this chapter we mentioned other appliances or tools that could be shared. We now show two different vacuum cleaners that are designed for communal use or ownership:

The first of these (see Figure 36) was developed by Mohammed Azali Bin Abdul Rahim (a postgraduate student from Malaysia) at Manchester. Like the lawn mower designed for shared use, this vacuum cleaner is also unusually sturdy to prevent damage through mishandling. As in the grass mower mentioned above, the motor, motor housing and wheels are a plug-in/plug-out component; in cases of motor malfunction, "downtime" can be reduced to a few minutes while the new motor unit is attached. The wheels are outsize and have heavy rubber treads so that the unit will not damage carpeting, and will easily wheel back to the storage facility. The central box area contains an outsize dust bag that is easily replaced, as well as all attachments.

The unit is primarily designed to be shared by six to eight apartment dwellers in one building. It is under consideration by the London County Council for use in County Council housing in the home counties, in and around London. This would widen its use since it could then also be shared by people living in six or eight separate small houses. (By the way, this is not a restrictive system, poised like a dagger at the throat of "Free Enterprise"; anyone who still gets his jollies out of owning his very own vacuum cleaner may of course buy one, in addition to the ones supplied by the local housing authority.)

The second shared vacuum cleaner we show (Figures 37 and 38) was developed by Mike Whalley and bears a greater superficial resemblance to present-day vacuum cleaners. Unlike the earlier unit, it is not modular in concept. Its features are:

1 It can be used as a canister vacuum for carpets, curtains and upholstery.

2 It can also be used upright for greater maneuverability and speed over large floor areas.

3 The head contains regulators for different types of pile carpets.

4 There is no beater on the agitator; instead, stiff brushes make it suitable for use on linoleum and other floor coverings.

5 It is, again, quite sturdy (compared to standard vacuum cleaners). The vacuuming head is surrounded by a heavy rubber guard.

6 The outside disposable bags are easily removable.

7 When a bag is full, the motor cuts out and a warning light blinks.

8 It is returned to a special storage facility that visually indicates in what house or apartment it is currently being used when not in place.

9 The machine is comparatively lightweight. The central storage facility can also serve as a recharge location for battery-driven, rechargeable models.

Another appliance that might easily be shared by both apartment dwellers and homeowners in a neighborhood is a community freezer with individually lockable compartments.

In the upright unit shown in Figure 39, each compartment has a capacity of 2.5 cubic feet (71.8 liters) and will hold approximately 48 pounds (21.8 Kg.) of frozen food. The inner doors have individual locks, while the major insulation door is opened by a master key. The unit is modular and will connect with other units horizontally. Each vertical bank of five individual compartments is 69 inches tall, 24 inches wide and 25⅛ inches deep (from wall to front-face of outer door).

A list of devices that might be shared, or owned by a central service facility within a small area, or leased through a service organization, could be endless. Snow-blowers are most necessary tools in areas such as northern Ontario, Minnesota and upper New York State. They are another type of tool that could be used cooperatively.

Until the inevitable emergence of communal, decentralized day-care centers, large play equipment (swings, slides, jungle gyms, round-

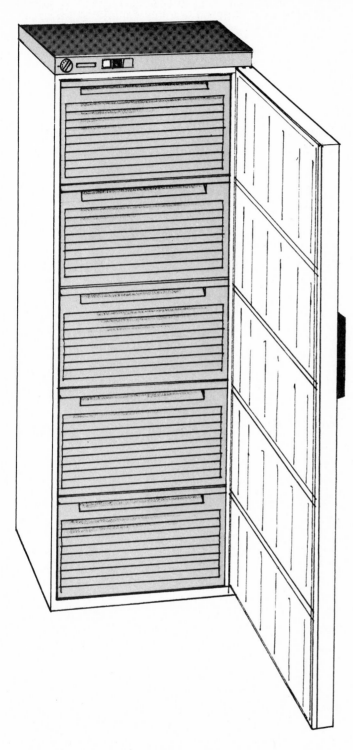

39 Community freezer with individual locked compartments

abouts, etc.) should also be communally and co-operatively bought and used.

Smaller electrical appliances such as mixers, blenders, sewing machines, as well as electric tools—circular saws, electric sanders, drills, saber saws, etc.—often lie unused and quiescent around the house for long time periods. These could be owned instead by a community service shop and used on a check-out basis. Even though a commercial start has already been made and some tools can be leased or rented, including rug shampooers and floor polishers that can be rented from neighborhood supermarkets, our whole concept of ownership needs re-examination. Our point is that many tools for kitchen, workshop and lawn care could be declassified as individual possessions and owned cooperatively by small groups.

The dividing line between tangible and intangible products and services sometimes appears blurred. While we were in the Midwest, we developed, designed and built with a group of students a neighborhood playground in a low-income area. Naturally all the design work was done in close consultation with the children who would eventually play there, as well as with their parents. The playground and the various play structures were built with the help of fathers, mothers and children on a flat lot 85 by 125 feet. When it was finally completed, including grassy areas, planting of shade trees (contributed by tree farms in the area) and with benches for elderly people, it was well worked out, logically designed and beautiful indeed. It had only one drawback: Children did not play in it! Subsequent meetings and discussions with parents and children showed that many of the mothers were unwilling to let their children go into an unsupervised play area alone. This direct feedback from the end-users of the design led to more community action.

As the playground was completely flat, and located in a midwestern state so flat as to appear virtually concave, we contracted for a neighbor with a bulldozer to create a mound, 4 feet high and roughly 12 by 15 feet, in the very center of the playground. On the hill we built a sort of double-insulated greenhouse (made of used angle iron and "culled" glass). The sides of the hill were planted with grass and provided excellent sledding in the winter. A neighborhood block party —complete with spaghetti and meatballs and a local music group—raised enough money to buy four used washing machines and a used dryer, all of which were installed in the greenhouse.

From here on, the playground was in constant use, serving the neighborhood in several socially valuable ways. The greenhouse became a natural center where mothers would gather, exchange community gossip, do their laundry, and at the same time be able to supervise their children at play. There were two further spin-offs: A neighborhood bulletin board soon made its appearance in the greenhouse launderette; and the use of a neighborhood laundromat (demanding exorbitant prices from this captive audience of slum dwellers) declined until it finally went out of business.

From this neighborhood sharing experience with its built-in social-encounter opportunities, a new motto might emerge for designers: *If you design a neighborhood playground, don't forget the washing machines!*

A good example of a shared intangible service deals with the maintenance of Jaguar automobiles. In the early 1960s a group of Jaguar owners in Buffalo, New York, banded together to solve their common need. Since Jaguars are unusually temperamental automobiles, needing constant maintenance, idling adjustment and minor repairs; since Jaguar owners (at least in Buffalo, and at that time) were not inclined to get their own hands dirty; and since the local service shops left a great deal to be desired, a group of about sixty people hired a highly qualified Jaguar mechanic from Great Britain and helped him resettle in their city, with the understanding that they would get his exclusive services. Nearly fifteen years later, this arrangement is still working.

This somewhat elitist example can be broadened to meet the more down-to-earth needs of the majority of the people living in neighborhoods:

We suggest the development of neighborhood skill banks. These might work on a cooperative basis without remuneration (unlike the Jaguar example given above) or on a small token-fee basis. Men and women with ability in such diverse areas as carpentry, plumbing, appliance repair, gardening, electrical work, baby-sitting and childcare, paramedical and parasociological work, accounting, income-tax advice, paralegal and real-estate counseling, and much more could place a few hours of their evening time at the disposal of the community. Eventually other needs—for supervised physical-exercise programs, foreign language skills, photographic darkroom services, lessons on musical instruments, etc.—may well emerge.

Some of these changes have already occurred in communities across North America. Their arrival is to be welcomed as a step toward greater neighborhood autonomy and a more decentralized way of life.

3

☯Why Not Build a Piano out of Used B-52 Wing Struts?

They [the instruction manuals] were full of errors, ambiguities, omissions and information so completely screwed up you had to read them six times to make any sense out of them. . . . They were really "Spectator Manuals." Implicit in every line is the idea that "Here is the machine, isolated in time and space from everything else in the universe. It has no relationship to you, you have no relationship to it, other than to turn certain switches, maintain voltage levels, check for error conditions . . ." and so on.

—Robert M. Pirsig, *Zen and the Art of Motorcycle Maintenance*

During World War II, Harry Rhodes spent most of his off-duty time designing and building a piano made from scrap airplane parts. Today Rhodes is better known as the innovator of the Fender-Rhodes electric piano. Seldom can a do-it-yourself job, combined with the recycling of scrap materials, have been carried quite that far or to so spectacular an end result. Instructions for building a piano in this way were printed in an armed-forces pamphlet and distributed to U.S. bases overseas.[1]

In this and the following chapter, we will deal with do-it-yourself approaches based on material available in kit form. We feel that building something yourself has basic advantages that transcend financial savings. First of all, it helps dispel the sense of alienation between object and user that has become unusually strong over the last two decades. People purchase gadgets, tools and artifacts, but once the purchase is made, the kick frequently goes out of the object, and the user feels estranged and sometimes even threatened by the high-technology appliance he or she has bought. Building things from a kit provides an understanding of how it all fits together and, to some degree, how it operates; hence, there is greater ability to cope with malfunction diagnosis, repair of faults and everyday maintenance. What you have put together yourself, you do not fear.

Secondly, it helps combat the loss of quality control that has occurred in most technologically advanced countries due partly to a spectacular increase in resentment of the product and alienation from it on the part of workers who have to work on assembly lines. Building it yourself, you are directly in charge of quality control.

In addition, building something yourself provides much greater satisfaction. By modifying the kit, you have the possibility of personalizing and even developing a one-of-a-kind object. Finally, one can transcend the kit itself and develop entirely new constructs.

Once you decide to build things from kits, you will find that your range of choices in the marketplace broadens magically. In some areas entire classes of objects are *only* available in kit form. The building process itself offers a whole new learning experience in terms of mechanical understanding, the development of motor skills and familiarity with new tools and concepts. And, of course, there is the benefit of cost reduction in building it yourself.

People who contemplate building something from a kit tend to consider hi-fi equipment first. But it is instructive to realize what a broad range of products is available in kit form for do-it-yourselfers and what the implications of so broad a

35

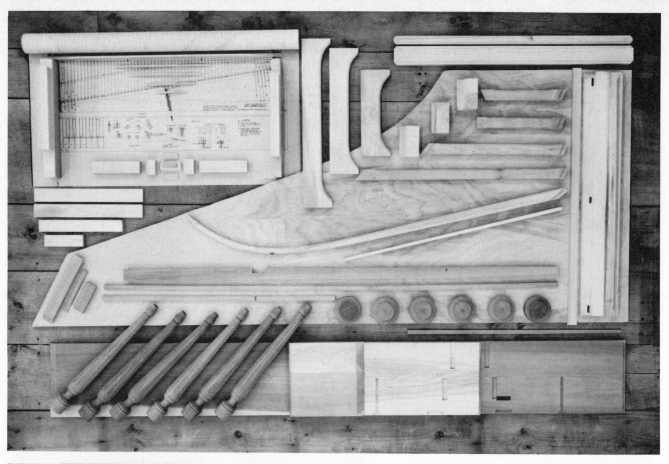

40 Harpsichord kit parts. This particular kit, from Frank Hubbard Harpsichords of Waltham, Mass., is one of the finest currently available.
Courtesy Frank Hubbard Harpsichords

41 A harpsichord, built from kit form by Dr. James Rubin of Rochester, Minn. This is a Burton harpsichord, again an unusually fine instrument, which introduced a Delrin jack and is hence less affected by temperature and humidity changes. This is probably the only major twentieth-century innovation in harpsichord building.
Courtesy Burton Harpsichords

range of products are or could well be.

The majority of some of the most exacting devices we use, such as harpsichords, top-of-the-line classical guitars, pianofortes and working replicas of Medieval lutes, come in kit form only. (See Figures 40 through 42.) There are good reasons for this. Whereas a conventional piano needs to be tuned only once every two or three years, a harpsichord should ideally be tuned before each use and be "voiced" occasionally. It would be not only costly but impossible to have a tuner come in every time you feel like playing, and you cannot tune a harpsichord (or voice it) unless you understand how it works. So the learning aspect of the building process is a prerequisite to being able to use the instrument.

Secondly, there are enormous savings in building one's own harpsichord instead of trying to buy one already assembled. In addition, although the pioneering work by people like Frank Hubbard, Dolmetch and others encompasses the building of the complete instrument, building instructions are usually open-ended enough to permit

42 A fortepiano by Frank Hubbard. The kit is shown partially assembled.
Photo Herb Gallagher. Courtesy Frank Hubbard Harpsichords

the user to experiment with the sound characteristics of other woods than the ones specified.

Not only exacting but also sophisticated precision objects are available on the do-it-yourself market. A recent advertisement for a Cronus model 2K electronic digital stopwatch promises accuracy of one second in one hundred thousand and offers this kit at a 45 percent savings over the assembled chronometer.[2] Some electronic music synthesizers are available in kit form at a fraction of the cost of Moog synthesizers. (See Figures 43 and 44.)

In this regard it is interesting to note that the majority of the breakthroughs in electronic music-synthesizer design have been made by people starting from kits. Heathkit (the largest maker of electronic equipment in kit form) lists among others the following items in a recent catalog: electronic digital alarm clocks; thermometers and weather stations; calculators; AM-FM digital alarm clock radios; amplifiers; four-channel receivers; tuners;

eight-track cartridge and Dolbyized cassette decks, stereo microphone mixers and reel-to-reel stereo decks, speakers and turntables; radio direction finders; digital depth sounders; foghorn hailers; dual-range fish-spotting sounders and other boating accessories; about thirty electric and electronic tune-up tools and diagnostic devices; aircraft and automotive accessories; walkie-talkies and CB gear; color organs and metal detectors; short-wave and ham radio equipment, electronic organs and home-protection systems; trash compactors; digital and analog power supplies; TV antennas; ultrasonic cleaners; electronic air purifiers; microwave ovens and much else.[3]

Among the eight or nine TV sets offered, their twenty-five-inch digital color TV is probably the most spectacular. It may be the most spectacular ever made, as nothing of that size and with those features is currently available as a manufactured item. Again, the most advanced unit is available in kit form *only*.

In the same Heathkit catalog, some of the items are available both in kit form or as assembled products. The price savings to the do-it-yourselfer range from 28 percent to 48 percent.

43 A modular electronic music synthesizer built from a series of subassemblies. The modules are made by PAIA Electronics, Inc., in Oklahoma City, who deserve commendation for their unusually detailed kit-building manuals.
Courtesy PAIA Electronics, Inc.

44 "Gnome" micro-synthesizer kit by PAIA Electronics, Inc.
Courtesy PAIA Electronics, Inc.

•

As an experiment, we tried to get someone without any experience in electronics, building kits, or even any prior experience in following instruction manuals to build a TV set. Sara Hennessey built a twelve-inch black-and-white portable TV set from a kit supplied by the Heath Company, without any guidance. It works as well as any TV set built from a kit, and because of worker-alienation factors, probably marginally better than a factory-made TV set. The only difference between Sara's endeavor and that of someone with kit-building experience was one of time.

Besides exacting and technologically precise products, many large constructs are also available in kit form. These range from greenhouses, garages and storage sheds, through geodesic domes, summer cabins, prefab homes, the completion of "pre-built" houses ("factory housing") and vacation cabins, to "second homes."

Some of the equipment used at the very edge of human survival comes in kit form only: most

Arctic and desert expeditionary tents, sleeping bags that operate at −50°F., backpacks, etc. Also available in kit form are sailplanes, gliders, small airplanes, helicopters, gyrocopters, the Bede one-man jet, canoes, fold-boats, kayaks, fishing boats, sailboats and small yachts. In addition, there are kits for such unlikely objects as dune buggies, sitars and Irish harps, houseboats, antique rifles, grandfather clocks, x-ray fluoroscopes, hovercraft, horse corrals, voltage regulators, celestial telescopes, pollution testers, yurts and biofeedback generators.

The learning aspect of building things from kits is probably best illustrated by the Fischer Technik electronic and construction sets for children (Figure 49). These sets, made and widely used in Germany, are available in this country through import toy shops such as F.A.O. Schwarz in New York City. Children attain a high level of understanding motors, drive shafts, statistics and electromechanics (switching and control). Fischer's electronic add-on kits provide in addition a basic

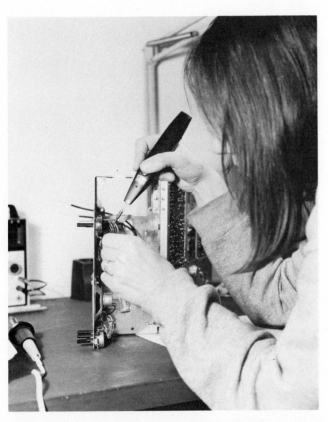

45 **Steps in the assembly of a Heathkit portable battery-powered TV. Sara Hennessey is shown soldering wires to the volume control.**

46 **Finishing installation of the two printed circuit boards**

knowledge of light, sound, signal and noise, steering and regulation, heat generation, etc.

Braun, of West Germany, has developed a series of electronic kits marketed under the name "Lectron" by the Raytheon Education Company of Waltham, Massachusetts, and distributed through Sears, Roebuck and Company (see Figure 50). The basic set consists of see-through transistorized blocks that connect magnetically and form working models of light-modulated electronic organs, automatic night-light control circuits, amplifier circuits, feedback circuits, moisture indicators and much else.

This whole concept—using do-it-yourself build-it-yourself approaches to learn—recalls the Chinese proverb:

> I hear and I forget
> I see and I remember
> I *do* and I *understand*.[4]

Sinclair Radionics Ltd. of St. Ives, Huntingdonshire, England, has for a number of years developed some of the most compact electronic devices for the do-it-yourself market as well as for regular commercial assembly and sales. Their most recent contribution is the Sinclair Project 80 (Figure 51), a high-fidelity stereo kit that can be purchased as several ultrasmall modular components: preamplifier and control unit; FM tuner; stereo decoder for FM stereo reception; active filter unit; power-supply modules and two power-amplifier options. These units measure one and a half by two inches at variable lengths and, being modular, can be interwired into many different configurations. This gives the do-it-yourselfer enormous decision-making powers and control over the size, appearance and materials used in the final, completed hi-fi system. We put the set together and mounted it in a linear control configuration on a single sheet of bent Plexiglas (as shown in Figure 52).

Sinclair has also developed a series of thin-line pocket calculators, beginning with the introduction of the "Sinclair Executive" which is less than half an inch thick.[5] This unit has been succeeded by even more compact calculators, several

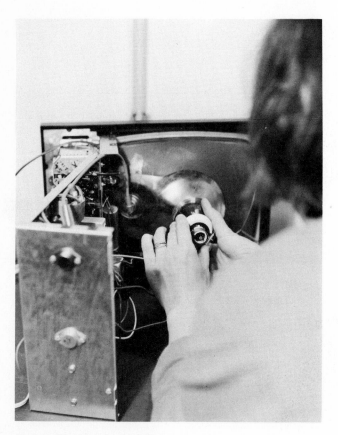

47   Installing the picture tube

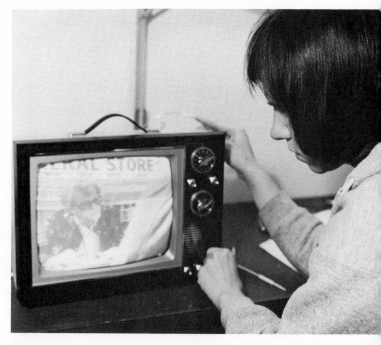

48   Adjusting the image on the completed set

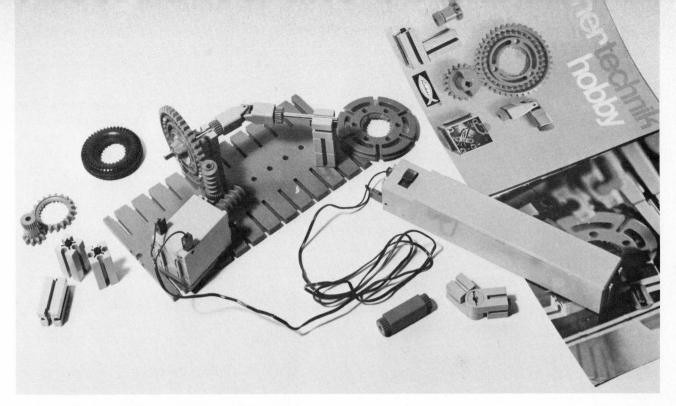

49 Electromechanical construction set by Fischer Technik

50 Electronic learning kit by "Lectron," distributed by Sears and made by Braun of West Germany

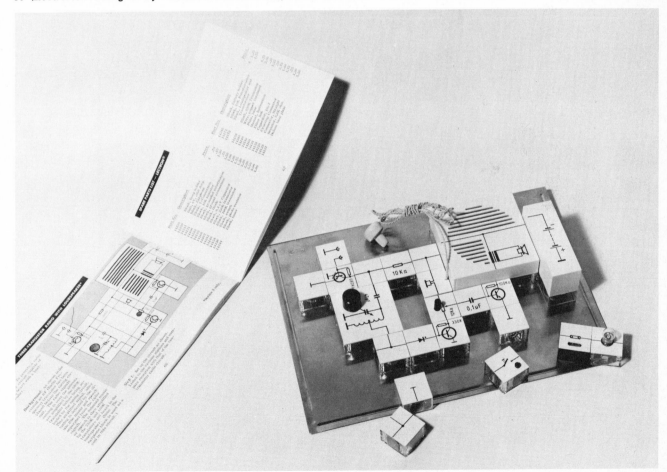

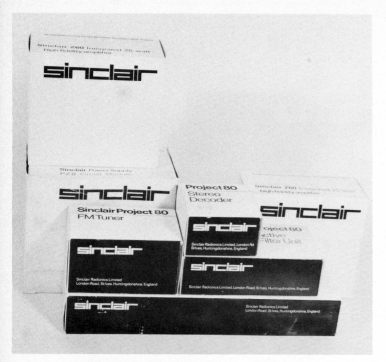

51 Sinclair Project 80 kit. This is nearly unique in being a kit of modules as opposed to a kit of components. All modules arrive in logical containers that are as free from frills as the contents themselves.

of which now come as do-it-yourself kits. Their latest (January 1976) market offering is the "Black Watch" kit for a digital LED wristwatch with calendar. We mention the Sinclair story as an example of microminiaturized chip technology for the do-it-yourself market.

◑ The most significant point about broadening the range of objects available in do-it-yourself form is that, once designed for assembly beyond the factory gate, products would change in kind, material and design. A simple example will suffice: A potter's kick wheel can weigh as much as 240 pounds. Primarily to reduce shipping costs, Pacifica Woodcrafts of Blaine, Washington, has developed a knockdown potter's wheel in which the standard flywheel (by necessity heavy metal or concrete) has been replaced with a cast fiber-glass flywheel that can be filled with pebbles, sand or water to regulate the weight. Empty, it weighs only five pounds (see Figure 53).

52 One possible arrangement of many alternatives

This concept of changing things because of do-it-yourself considerations has been applied by some of our students to everyday household objects (see Figure 54). Figure 55 shows the sectional views of a standard centrifugal spin dryer and the unit as it might appear when redesigned for less waste of energy and materials, lighter shipping weight and home assembly. At the base of conventional spin dryers is a thirty-eight-pound iron casting (to keep it from moving around). At the bottom of this casting are four wheels (so that it can be moved around more easily), and on each of the four wheels is a brake (to keep it from moving around). Between the base and the dryer above it are four nasty metal springs (to keep it moving, but not too much). The result is that the conventional spin dryer slowly moves across the floor with a flapping, hopping motion like a huge maimed bird. After a few years the continuing vibration causes cracks in the base, which finally shears.

The redesigned version by Mark Hofton substi-

tutes a plastic cushion that can be filled with water to gain forty pounds of weight when in use. This cushion in turn sits on an eight-inch-thick foam pillow, which reduces all vibration and noise, establishes enough friction to keep the unit from jiggling across the floor, but permits it to be pushed in any direction with a firm lateral push.

Joseph Rosenbloom's book *Kits & Plans,* a listing of manufacturers only, and by now more than three years out of date, runs to nearly three hundred pages and has thousands of entries.[6] But while the list of things now available in kits is incredibly long, some of Rosenbloom's entries are quite baroque, since few ordinary household devices exist in kit form.

In an earlier book we showed that the Luxo desk or drafting lamp, then costing nearly $30, was also available as a do-it-yourself kit from Allied Radio Corporation in Chicago, Illinois, for $9.[7] To "build" the lamp, three screws had to be tightened, six feet of wiring had to be snaked

53 The Pacifica potter's wheel

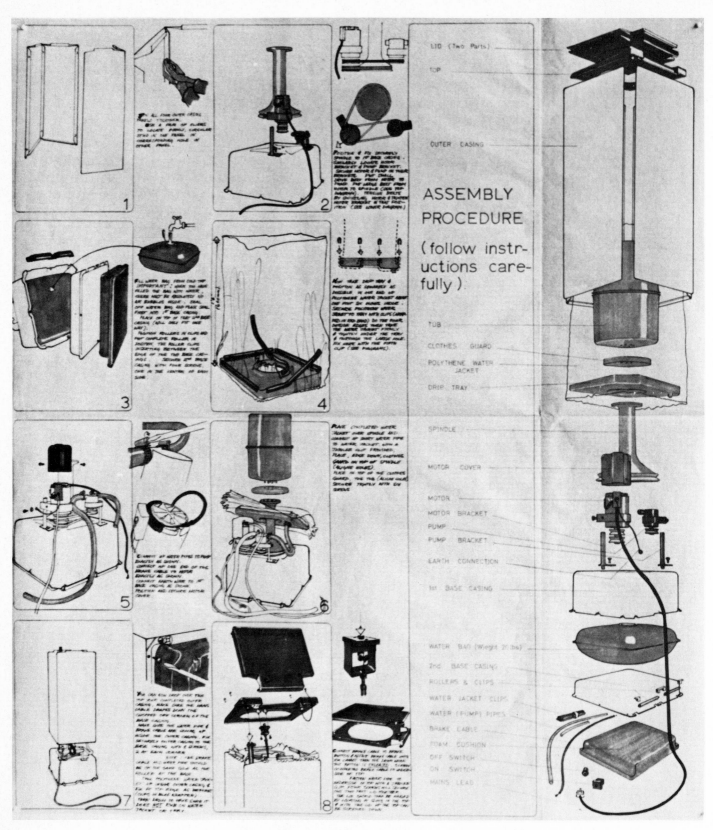

54 Redesign of spin-dryer for lighter weight and kit-building, by Mark Hofton, a postgraduate student at Manchester Polytechnic
Photo Kokon Chung. Courtesy *Design Magazine*, England

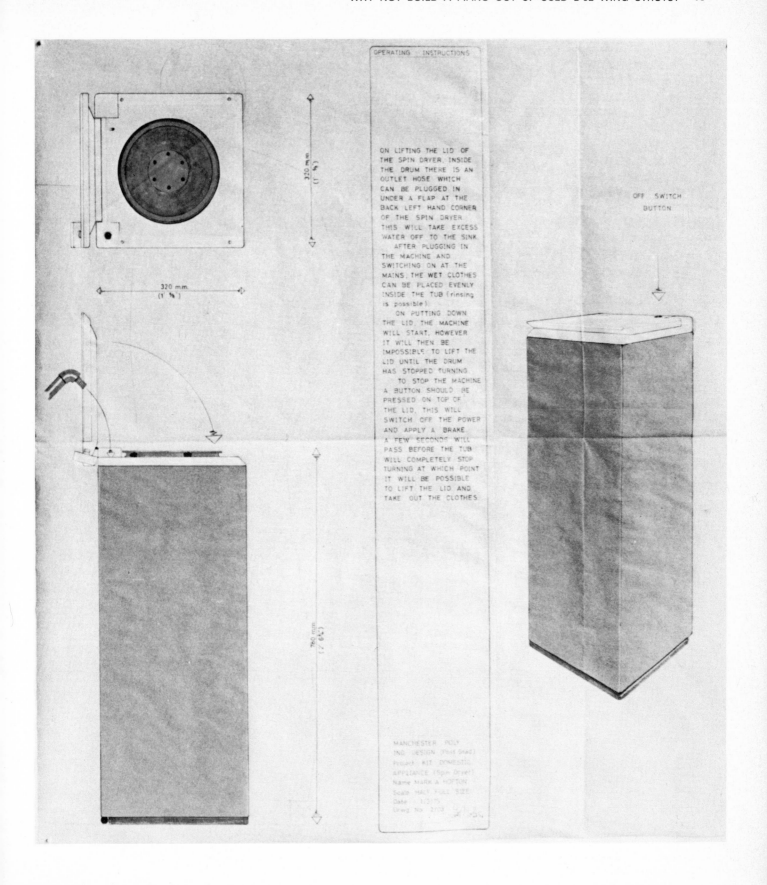

ON LIFTING THE LID OF
THE SPIN DRYER, INSIDE
THE DRUM THERE IS AN
OUTLET HOSE WHICH
CAN BE PLUGGED IN
UNDER A FLAP AT THE
BACK LEFT HAND CORNER
OF THE SPIN DRYER.
THIS WILL TAKE EXCESS
WATER OFF TO THE SINK.
  AFTER PLUGGING IN
THE MACHINE AND
SWITCHING ON AT THE
MAINS, THE WET CLOTHES
CAN BE PLACED EVENLY
INSIDE THE TUB (rinsing
is possible).
  ON PUTTING DOWN
THE LID, THE MACHINE
WILL START, HOWEVER
IT WILL THEN BE
IMPOSSIBLE TO LIFT THE
LID UNTIL THE DRUM
HAS STOPPED TURNING.
  TO STOP THE MACHINE
A BUTTON SHOULD BE
PRESSED ON TOP OF
THE LID, THIS WILL
SWITCH OFF THE POWER
AND APPLY A BRAKE.
A FEW SECONDS WILL
PASS BEFORE THE TUB
WILL COMPLETELY STOP
TURNING AT WHICH POINT
IT WILL BE POSSIBLE
TO LIFT THE LID AND
TAKE OUT THE CLOTHES.

320 mm
(1 ¾")

320 mm.
(1 ¾")

760 mm
(2 6¼")

OFF SWITCH
BUTTON

MANCHESTER POLY
IND. DESIGN (First Year)
Project KIT DOMESTIC
APPLIANCE (Spin Dryer)
Name MARK A HOPTON
Scale HALF FULL SIZE
Date 1/2/75
Drwg No 2/05

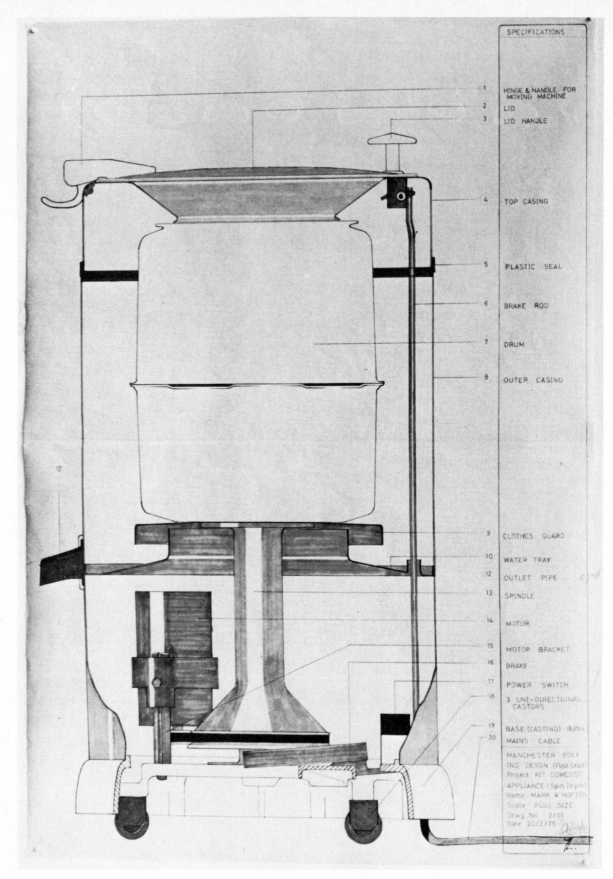

SPECIFICATIONS

1   HINGE & HANDLE FOR
    MOVING MACHINE

2   LID

3   LID HANDLE

4   TOP CASING

5   PLASTIC SEAL

6   BRAKE ROD

7   DRUM

8   OUTER CASING

9   CLOTHES GUARD

10 WATER TRAY

12 OUTLET PIPE

13 SPINDLE

14 MOTOR

15 MOTOR BRACKET

16 BRAKE

17 POWER SWITCH

18 3 UNI-DIRECTIONAL
    CASTORS

19 BASE (CASTING) 16lbs

20 MAINS CABLE

MANCHESTER POLY
IND. DESIGN (Post Grad)
Project: KIT DOMESTIC
APPLIANCE (Spin Dryer)
Name: MARK A HOFTON
Scale: FULL SIZE
Drwg No.: 2/01
Date: 20/2/75

55 Comparison of typical spin-dryer and redesigned dryer on the right (by Mark Hofton)
Photo Kokon Chung. Courtesy *Design Magazine*, England

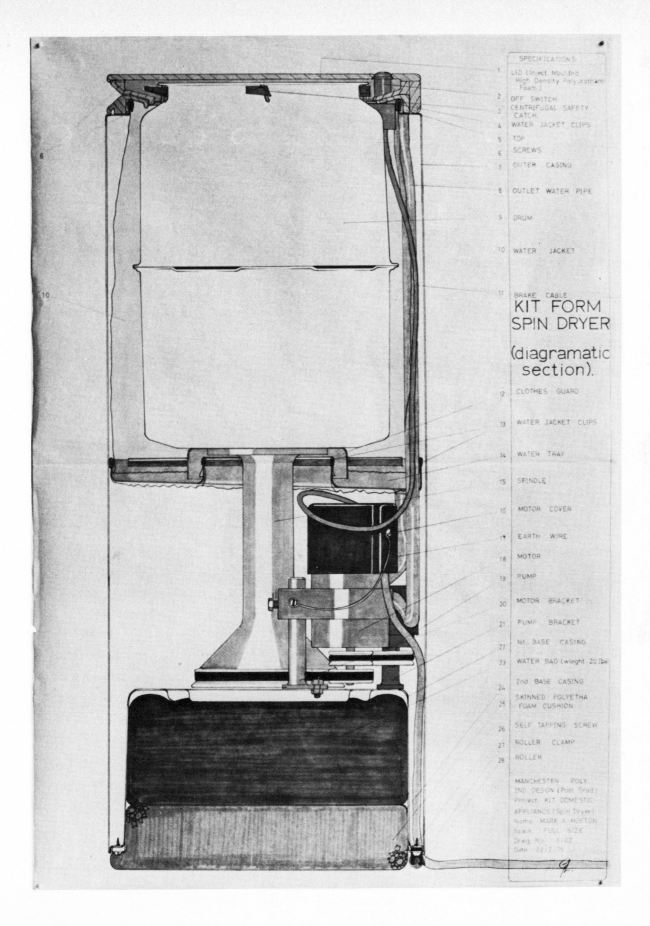

SPECIFICATIONS

KIT FORM
SPIN DRYER

(diagramatic section).

through one of the arms, one nut secured, and a plug attached. All in all, ten to fifteen minutes' work and a savings of $21.00. There seems no question that a person capable of so simple a task could equally well assemble a box fan, a steam iron, a humidifier, a hair dryer, toaster, percolator, electric frying pan or electric clock. Several weeks of evening kit-building would be needed to get into more advanced objects such as washers, clothes dryers, air purifiers, air conditioners, sewing machines, portable power generators, photographic enlargers, projectors, baby carriages or what-have-you. Unfortunately none of these items are available in kit form as yet.

This is not to suggest that everyone should build from kits. Many people, after a day's hard work on an assembly line, would find spending the evening putting a vacuum cleaner together somewhat less than enchanting. Others may lack the mechanical ability or the simple motor skills to work effectively with their hands. The point is that for an undeterminedly large part of the population that can and would like to build household appliances, this choice does not presently exist in the market.

Some words of caution might be added: One of the worst aspects of do-it-yourself work is the poorly conceived, badly written and badly illustrated instruction manual (see the quotes introducing this and the following chapter). Nor are all do-it-yourself kits well designed or of good quality. Usually kit producers (at least the producers of quality kits) are quite willing to share with you past results of their users, as well as to pass along to you the names of satisfied customers in or near your community. Obviously, kits that have been made successfully by many people are preferable to those that have produced a string of failures. Failure is probably the greatest discouragement to a beginner, for if the first try turns into a dreary collection of VU meters, mounting brackets, toggle switches, I.F. strips and sheared bolts rattling around the bottom of the wardrobe, another possible enthusiast has been stopped cold.

But in spite of the cautions we have just expressed, it is our sincere belief that people are more capable of doing things for themselves than they believe. Especially in North America, a nation of tinkerers, where many of us spent our teens working with cars and sound equipment, the total ability of the population to handle self-built projects is very high. Nor should we discount the enormous educational changes among young people today. People have more education (often gained through television and radio rather than the schoolroom) and are often trained in mechanical skills in the military or the Peace Corps or VISTA. Finally, an increasing part of the population is engaged in clerical work of one sort or another. The natural human need to accomplish something with one's hands is very strong in this group. The recent economic recession has also made many willing to try new ways of reducing their expenditures—expenditures that frequently run high because of exorbitant repair, service and maintenance costs.

How this reservoir of knowledge and skills can be tapped will be discussed in the next chapter.

4

◐ Why You Can't Build a Lemon Alone

While depressing flaps A, B, C and D, thread Flotation-Deck-Supports 11A and 11B through Pool-Edge-Siding "K" and at the same time force the entire sub-assembly into Edge-Receptor-Slit 22 (Secure with nuts at 4A, 4B, 4C, 4D and 4E). . . .

—From an instruction sheet explaining how to build an above-ground swimming pool for dolls

Between 1966 and February 1975 the automobile industry recalled 45,700,000 automobiles for inspection or repair.[1] Many of the troubles are the result of worker dissatisfaction and, more importantly, an alienation between the worker and the product that is even more deeply rooted than the alienation between user and object, discussed in the previous chapter. The dreary sameness of assembly-line procedures, the unremitting dependence by man on the timing and rhythm of the machine process, the horrifying monotony of workday and workweek, are some of the primary causes of this condition.

Possibly it all began with efficiency experts, time-and-motion studies, Taylorism and the "findings" arising from that curious collection of claptrap, superstitious reductionism and pious ineptness that can best be referred to laughingly as "behavioral psychology." Almost prophetically Karl Marx said: "They want production to be limited to 'useful things,' but forget that the production of too many 'useful' things will result in too many 'useless people.' "[2]

It must be noted that we are mainly concerned here with mismatching and assembly-line errors, or deliberate assembly-line sabotage. Hence cars malfunctioning for *engineering* reasons cannot concern us here. With respect to the latter problem, one thinks, for example, of General Motors' famous Vega where at first, in April 1972, 130,000 Vegas had to be recalled; with a second recall in May 1972, 86 percent of all Vegas were brought back; and finally during the third recall in July 1972, 90 percent (nine out of ten!) of all Vegas produced had been recalled because of engineering defects.[3]

Design flaws were responsible for defects in about two-thirds of the 52 million cars and trucks recalled since 1966 to correct safety problems, according to federal safety officials. A further one-third were caused by assembly-line work and sabotage connected with work alienation. Although both Ford and Chrysler are unwilling to reveal the annual cost of recalling and repairing defective cars and trucks, Thomas A. Murphy, GM Board Chairman, told his shareholders in May that the company spent $5.7 million in 1975 alone. Just the certified *postage* to owners of the 770,000 vehicles most recently recalled cost GM and Ford approximately $400,000.[4]

Worker alienation and worker apathy, however, lead to even more serious consequences—such as "Monday" or "Friday" cars (automobiles left partially unfinished, either because of an anticipated weekend or a hangover from the last one). Then there is deliberate sabotage, such as welding several loose crown nuts into a metal pocket below the glove compartment to insure an undiagnosable permanent rattle.

There is also the problem of absenteeism, either deliberate or caused by psychosomatic illnesses. Such absenteeism frequently runs to over 22 percent of assembly-line workers. Other social pathology, such as alcoholism, drug addiction and even suicide, has been linked to working conditions on assembly lines.

This is by no means a phenomenon exclusive to the United States. Identical, or nearly identical, findings come from British Leyland, Renault and Citroën in France; Fiat in Italy and South America; Volkswagen-Audi-Porsche and BMW in Germany; Volvo and Saab in Sweden; and were it not for censorship, one would imagine nearly identical findings from the Moskvitch factory in Russia, Seagull in Peking and Tatra in Czechoslovakia. (We'll talk about Japan later.) Nor do we mean to suggest that this alienation is restricted to automotive workers; they merely provide the most widely documented examples.

Approximately eight years ago the automobile concern Volvo AB, in Gothenburg, Sweden, felt that with the rejection rate and return rate of automobiles, ambulances, buses and trucks steadily climbing, it would be just plain good management to find out how the troubles of the assembly line could be alleviated. In a dramatic move it was decided to build a new plant outside the city that would do away with assembly lines entirely. Instead, cars would be assembled by teams of sixty to ninety workers within their own assembly space, and with the team determining work rate. It was furthermore decided (in direct consultation with the workers) that all the different tasks within a team, including chairmanship of the team, would rotate. This would eliminate boredom and job fatigue and give each worker a chance to perform all of the functions that go into the making of an automobile. He or she would derive greater job satisfaction from having *de facto* put together entire automobiles "alone" after a time.

The program, augmented with many refinements over the last few years, works exceedingly well. Teams have effective control over their work schedules and complete autonomy in their recreation areas adjoining the assembly room. There is also full discretion over break times, lunch, etc.

Suggestions from the factory floor are not only welcome but eagerly sought; they are rewarded not with incentive pay but (by the workers' own preference) with extra holiday time.

In solving any problem in a new and dramatic way, dozens of new difficulties are encountered. Two design problems will suffice.

In trying to assure complete job rotation within each team, it was found that some component parts of trucks and buses were too heavy for the majority of women to handle. Meanwhile, male guestworkers from Turkey, Spain, Yugoslavia and Italy were uncomfortable with the task of upholstering car seats, which they equated with sewing and thus considered "unmanly."

To solve the first problem, handling machinery was beefed up and the weight of individual components reduced to make job performance by women employees less difficult for them. As for the men, the entire upholstery process had to be redesigned in order to "desex" it and not offend the cultural self-image of foreign male workers.

We worked as senior design consultants for Volvo during some of this time and are aware of the startling results in improved worker morale, the nearly complete withering away of absenteeism and the total disappearance of work sabotage. But further strides must and will be made in this work-enrichment process. Soon a small metal plaque inside the door of each Volvo made at the Kalmar plant will inform the owner that it was built by, say, the "Robinson Crusoe Team" and is, for instance, their 109th automobile. Should the owner then have problems with the car, he or she will be put directly in touch with the team that built the car, rather than with a faceless corporate structure. This will tie user and producer into a much closer relationship.

Most recently, some of the workers at Volvo have even begun to ask: "If ninety of us can build a car in a room, then why not put that room in our village? Then we would not have to come to the factory at all." The surprising thing is that Volvo is in principle enthusiastic about this approach. However, the logistics of raw-materials handling and the problem of how to keep manufactured parts flowing to the assembly teams are such that the Swedish countryside is not ready to absorb a hundred or so decentralized assembly stations without suffering severe ecological damage.

It is, however, a portent of things to come.

Already teams frequently contain a husband, wife and grown daughter or son, and the network of personal bonds within each team is very strong. (On an *ad hoc* basis some decentralization is already occurring; a growing number of workers at Volvo and also at Saab, which is attempting to build a similar system, take subassemblies home for double pay to work on over the weekend.)

Early in 1975 a large group of automobile workers from the United States was taken to Sweden to work at Volvo and Saab. Their reactions, as reported in the international edition of *Time,* were almost totally negative and are best summed up by the response of one of the workers from General Motors who, when asked why he didn't like the Volvo assembly-team program, replied: *"You gotta think!"*[5] This answer paints an ugly picture of the extent to which the assembly line has dehumanized workers. Without belaboring social, political or psychological questions, it can at least be said that workers so repressed by the routine nature of the work process that they have trained themselves not to think about what they are doing will have frequent work injuries and accidents.

The financial results of Volvo's experimentation with work time and working conditions are instructive: While the labor cost per car has climbed, the rejection rate of cars has nose-dived so dramatically that marginal *de facto* savings are taking place. The important thing, however, is that in the midst of the world oil crisis Volvo is looking for alternatives to the internal-combustion engine, as well as making the transition to a new technology more bearable.

At the time of this writing (May 29, 1976) Volvo is installing a catalytic converter ten times more effective in reducing harmful pollutants and emissions than any other currently being used.[6]

In looking for alternative power sources, Volvo at first seems trapped by its own philosophy of car safety: Having decided some years ago to attempt to make the safest car in the world, Volvo soon realized that this would also mean keeping the automobile relatively large and heavy. As long as other massive cars like Cadillac, Chrysler, Mercedes Benz, etc., are on the road, a really safe car needs enough weight to withstand crashes. Present-day battery technology works only on short trips, with small, lightweight automobiles.

Björn Ortenheim of the Institute of Technology of Upsala University has perfected the "Silentia" battery-driven car. Primarily conceived as a "city car," the four-seater can cruise 130 kilometers per charge (80 miles) at a top speed of 70 kmh (45 mph). The car weighs 610 Kg (1350 pounds) and is 3.28 meters long (11 feet). Thirty percent of the expended energy is recouped when going downhill, decelerating or braking. It has four electric motors, one driving each wheel, two of which are auxiliary

56 How the Volvo bimetallic system might work (schematic)

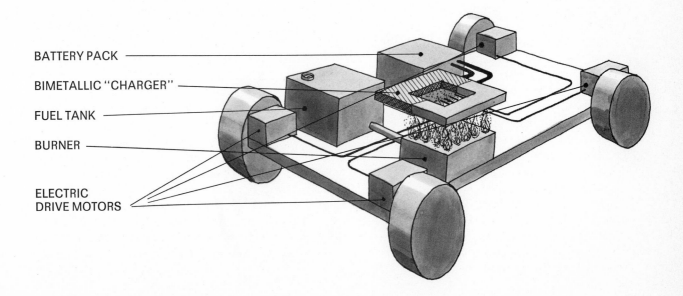

BATTERY PACK

BIMETALLIC "CHARGER"

FUEL TANK

BURNER

ELECTRIC DRIVE MOTORS

and used only during acceleration, saving a further 30 percent of energy when not in use during normal sustained driving. Power plant consists of ten batteries developing 7 KW (9.3 HP). Other features to reduce energy consumption include magnetic overdrive, automatic electromagnetic gearbox and a differential lock. A two-seater version can cruise at speeds of 90 Km/hr (56 mph).[7]

To recharge the Silentia and other electric vehicles, Thrige-Titan AS of Odense, Denmark, has developed an instant roadside recharger called "Parkelmeter." The unit has a built-in transformer, with six twenty-four-volt outlets complete with six-and-a-half-foot recharging cables that plug into the vehicles and a monitor light that goes off when each car has been recharged. The unit is coin-operated (like a parking meter) and, using household electric current, can be installed anywhere. In appearance and size, a Parkelmeter for six cars is somewhat smaller than six regular parking meters lined up side by side.[8] Volvo, which has recently acquired DAF of Holland, plans to develop the electric car through their smaller-bodied DAF division.

For their larger "safe" cars, Volvo has set up a small laboratory with a staff of six and the inventor Baltzar von Platen to develop a battery-powered, electrically driven car that will be independent of electric lines by carrying its own fuel plant and burner, which converts heat to electric power for recharging (see Figure 56). This operation will occur continuously while driving, thus eliminating the need for lengthy recharging stops.

Von Platen's invention exploits a discovery made a century and a half ago by the German physicist Johann Seebeck. The Seebeck effect can be demonstrated with a ring made from two lengths of different metals with their ends joined together. If one section is heated while the other is kept cool, an electric current will flow in the ring. Conversely, if a current is passed through the ring, one section will heat up and the other cool down. Von Platen first used the principle in another Swedish firm, Electrolux Refrigerator, in the early twenties.

Von Platen has obviously overcome the inefficiency of early thermoelectric devices. A typical car engine is about 30 percent efficient—the rest is wasted through heat. To better this, he has developed a material that can make thermoelectric devices about ten times more efficient than any presently known. Volvo has world rights to the device, and the paper *Expressen* surmises, on the basis of Volvo's contract, that they anticipate a sale

57 How a FIAT 127 might be built from a kit

of millions of cars on a worldwide scale. Besides the right to license other producers, the company has retained the right to produce units for nearly three million motor vehicles a year, more than 400,000 boat engines, over 30,000 industrial units and 500 aircraft engines. This means that, once the device is perfected, Volvo expects to produce ten times as many cars as their current annual production run of about 300,000.[9]

We have devoted so much space to the description of the Volvo for two reasons: First because of the inherent relevance of the subject matter, but, more importantly, because their teamwork approach is radically different from standard European and North American car-assembly methods. (Japanese auto makers put a padded room at the disposal of their workers which is furnished with stuffed dummies. These the workers are encouraged to beat up, strangle, kick, maul and maim. It is a desperate last-ditch effort to channel the assembly-line worker's hostilities and aggressions away from the product, boss, coworker or wife.)

An alternative solution to assembly-line production would be to make certain automobiles available in kits. If small automobiles like the Honda "Civic," the Volkswagen "Golf" ("Rabbit" in the U.S.), the Renault "5TL," the Chevrolet "Chevette" or the Fiat 127 (Figure 57) were available in kit form, many of today's assembly-line problems would disappear.

All parts for the assembly of mass-produced cars are built to upper and lower tolerance limits. Unavoidable mismatching and a buildup of sloppy fits or tolerances will occur among the thousands of parts that make up an automobile. Conceivably an overtoleranced hole, for instance, will be mated with an undertoleranced shaft, resulting in extreme vibration, wear and eventual breakdown. Conversely an undertoleranced hole mated with an overtoleranced shaft will probably be rammed together on the assembly line and fail to work properly, if at all.

Multiplying this possibility by the tens of thousands of parts that combine to make up a car, taken in conjunction with generally bad workmanship, dehumanizing work processes, warranties *specifically excluding certain trunk, hood and door alignments,* ill-prepared finishes, cursory inspection and quality control, you can see why the chances of your getting a "lemon" are moderately high. Add to this the fact that in 1972 (the last year for which such figures are available) most defective parts needed to be replaced rather than repaired. Spare parts are sold to dealers and then to the public at prices so inflated that it would cost more than $6400 to separately buy the parts for a $2500 Ford "Maverick."[10]

You can't build a lemon alone. Putting together an automobile from a kit, regardless of experience or training, you are better than several hundred hung-over assembly-line workers who hate their meaningless, mind-blinding, small, perfunctory tasks on the job, as well as the product itself.

In the early sixties some little kid wrote in to the then popular television program "You Asked for It." His request was to see if someone could assemble a car alone. The TV program, in order to establish credibility and prove that nothing had been preassembled, asked a local handyman to build a Hudson "Hornet" on top of a building.

The man chosen did the job alone and on top of the building in one week, and demonstrated the result by starting the engine and operating windshield wipers, lights, etc. While he had technical guidance (surely no more detailed than that available in any first-rate kit-building manual), he had no actual help. Nonetheless he completed the job without any errors except for breaking a windshield, which he had to replace. (Even the windshield might not have broken had the wind factor on top of the building not interfered with a single man picking up a large and bulky windshield alone.)

The fact is that the building of an entire car, even with limited time and in an ill-chosen environment, is perfectly feasible with no previous training. Furthermore, due to the economic recession and the high cost of auto maintenance, there is a growing number of people who themselves rebuild, replace, repair, service and maintain the mechanical and electrical components of their cars. With more and more fiber-glass conversion kits available for cars, especially Volkswagens, large numbers of people thus do their own interior fittings as well as body and coach work. In most cases the coach work is of a decidedly higher standard than that of mass-produced cars. This also holds true of body and engine conversions.

Thus a sizable public already exists who could build cars from kits. A service manual that has become a classic—"How to Keep Your Volkswagen

Alive: A Manual of Step-by-Step Procedures for the Complete Idiot"[11]—virtually enables one to build a Volkswagen from scratch right now, if only the parts or the entire car were available in kit form.

Making a large and complicated construct from a kit is no more difficult than, say, building a light dimmer from a kit available for $7.98. It is just more time-consuming. As with all good kits, one would first build a series of subassemblies and then plug them together. Building a car involves a product smaller than a geodesic dome, less complicated in its subassemblies than Heathkit's black-and-white TV set (built by Sara and shown in the last chapter), with much less exact tolerances than a celestial telescope and less stringent performance requirements than an expedition sleeping bag.

With all kits, if a part doesn't fit perfectly or function properly, the customer returns it, and it is replaced at no charge with a new and proper part. Our automotive kit builder then, if receiving a hood that doesn't quite make it, would keep returning it until the properly fitting part is obtained. This is a far different story from that of the normal assembly-line process, where the part must be "bashed to fit" because of time and working pressures, or the worker ignores it with a misplaced and optimistic reliance on inspection procedures further down the line, which are frequently superficial.

In the previous chapter we established that making consumer products available in kit form will radically alter materials, processes and assembly methods—eventually the design itself. We have also seen that making complex things available in kit form introduces the concept of subassemblies. Modularity is the result of subassemblies that are designed to plug together. Therefore complex kits rely on modular components that are compatible and, by extension, conducive to greater flexibility through the introduction of a variety of interchangeable modules, themselves available as kits.

◐ This variety of modular, interchangeable kits would lead to the introduction of a wide series of function packages. For instance, engine options would include one-, two- or four-cylinder gasoline, diesel, turbine, and electric engines and, eventually, alternative power sources. Frame options would allow for front-wheel drive, midengine positioning, rear-wheel or four-wheel drive, as well as leaf, independent strut or air suspension.

The transmission-package option might include your choice of standard, semiautomatic or automatic versions of fluid, magnetic, belt-driven or conventional gear-driven types.

A seating option, in addition to offering a broad choice of fabrics, colors and textures, would, more importantly, provide seat configurations such as front- or rearward-facing seats for passengers, horseshoe-shaped passenger couches, reclining seats, provisions for seating handicapped people in wheelchairs, seats on automatic swivel returns, and specialized interchangeable seats for children at various stages of their growth.

Fiber-glass body options would basically include packages for two- and four-door passenger cars, vans, campers and pickup trucks, station wagons and sports cars. Beyond that, even more specialized versions such as ambulances, dune buggies, fire engines, taxicabs and ultracompact city cars could be offered.

Other manufacturers would probably, in addition, begin to produce a myriad of other configurations to fit the framework of your kit. Conceivably, an endless variety of choices would result.

◐ Tage Schmidt is a friend and former student of ours at the Royal Academy of Fine Arts in Copenhagen. He is a writer concerned with technology and planning for a low-energy society. Having worked as a consultant for the Danish State Department of Commerce in the energy information field and with a multidisciplinary team to integrate bicycles into public transportation, he brings to his proposals an eighteen-year background as one of northern Europe's leading automobile-test drivers and journalist-critics. In addition to testing all new European and some American cars (from the smallest Fiats to Rolls-Royces and Ferraris), he has tested sports cars at speeds up to 150 mph (*sic*). His driving manuals are still used by Danish police driving instructors. All this, plus his strong recent position as a consumer advocate in design, lends great authority to his proposals.

In 1973 he suggested that a lightweight, eight-horsepower, thirty mph automobile, based on the prototype Citroën 2CV of 1936, might well be the most reasonable car to drive in the energy-starved and overcrowded world of the 1980s. (See Figures 58 through 66.)[12]

The proposal sounds wildly impractical at first. While the gasoline savings alone would amount to more than 20 percent (the car *averages* 54–59 miles per gallon), its light weight (it weighs only one-third as much as an average U.S. automobile) would provide further fuel savings and, more important, save precious, depletable energy and non-renewable resources in manufacture. Because the car is lightweight and slow, the number and seriousness of road accidents would decrease dramatically. Tage Schmidt goes further: He argues that if *all* small passenger cars and small trucks were based on the technical specifications of the 1948–54 Citroën 2CV, *direct* fuel consumption would fall by at least 20 percent, and still further gains in gasoline economy could easily be made. The car, in its original 1948 state, was unusually safe on icy roads (with no winter salting needed on roads, thus reducing both corrosion and the materials and energy spent on rust prevention and also eliminating another fresh-water pollution factor), and easy to keep going in deep snow. Tage Schmidt rightly argues that there is no need to accept the 1948 version, or its degenerate, modestly souped-up latter-day descendants. Allowing for simple minor present-state-of-the-art improvements, he manages to list dozens of impressive gains that could be made in energy savings, environmental quality, public health, emission controls without catalysts, and so forth.

Adolf Hitler's original "people's car" (Volkswagen) of 1938 owed much of its concept to the prototypal 1936 Citroën 2CV.

Today it is rare but not that unusual to come across a 2CV in Europe that is being driven for its thirty-seventh continuous year! Five or six of the original prototypes are still giving good service *now* after forty-one years of driving, with half a million miles or more on the odometer! The 1976 model sells for about $1600 (exclusive of taxes).

Tage Schmidt's plea on behalf of a more efficient low-energy car design may still seem far-out and barely feasible, but then our present situation is, ecologically and socially speaking, less feasible still.

We include Tage's ideas here to show that his vision could easily take shape if the proper option packages from one of our proposed kits were chosen. But eventually even his radical proposal could be improved upon (when the state of the art reaches that level) by unplugging the engine package and inserting a "solar-power" engine or some other kind as yet undreamed-of.

But one caution: The 2CV is an extremely haz-ardous vehicle to mingle in traffic with large, heavy and fast cars. Hence the entire Citroën 2CV scenario depends on *all* cars becoming similar transportation devices once again.

◐ Further changes in the design of kit automobiles will be forced into being by weight and shipping conditions. As we have seen in the previous chapter, the potter's kick wheel changed from a heavy concrete disk to a water-filled fiber-glass drum. In much the same way automobile bodies will be made of fiber glass or other impact-resistant plastics for greater ease in handling by a single kit builder, as well as for energy savings and lower shipping costs.

◐ This opens up two important new options. Fiber-glass bodies have much greater possibility for do-it-yourself repair after collisions. And secondly, with present research into the loose-bonding of plastics to make them biodegradable or biodegenerative (throwaway plastic beer cans that biodegrade have been in use in Sweden for over a year), car bodies can be built that will return to the soil when no longer needed. The body parts could be protected by a thin film screening out the ultraviolet rays of the sun; the film would be peeled off when the useful life of the body has ended.

◐ Again for shipping and handling reasons, the present-day complexly machined and heavy cast-iron engine block would have to become lighter. While lighter, smaller blocks are feasible, leading to a de-emphasis of horsepower, even turbine engines (smaller and lighter because of their high speed) might soon emerge as an alternative.

◐ Four small electric motors, each driving a wheel, are the smallest and lightest power package that can be put together. With present-day technology this might emerge as the most logical alternative option.

◐ Or—the electric motors could be powered by a flywheel, a Swiss system that has been used on buses in Europe and Africa since the early fifties and is now being used on trolleys for San Francisco built by Lockheed.[13]

Reaching back to the lightweight Citroën 2CV, it is useful to compare its seating arrangements with that of conventional cars. The upholstery and seat construction of conventional automobiles reflect the design and structural considerations of the upholstered-furniture industry of fifty years ago. Seats are heavy, massive, bulky, oversprung, underadjustable and reminiscent of first-class seats

58 1936 prototype for Citroën 2CV

59 First mass-produced Citroën 2CV, 1948

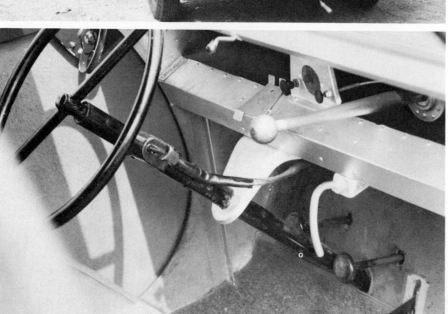

60 Dashboard showing horizontal stick shift

61 1976 version of Citroën 2CV

62 Citroën 2CV light van, 1976, produced in versions with different load capacities. The short version has a payload of approximately 550 pounds and runs at 47 miles to the U.S. gallon (remarkable for a truck!). The long version holds 1,050 pounds.

63 "Pierre," a Citroën 2CV, owned by Craig McArt

64 Rear view, with canvas roof rolled back

CITROËN

65 Side view

66 The cockpit, showing tubular aluminum seats with stretched canvas covers

67 **Interior shot, showing aluminum and canvas sling**

in a turn-of-the-century railroad car. This fact is attested to by those members of the "hippie" culture who prefer to sit in their living room on a van seat rather than an overstuffed couch. To their amazement, car and van seats are equally uncomfortable.

Compare this with the Citroën 2CV. The seating resembles lawn chairs or beach furniture, being made of bent aluminum tubes with stretched canvas slings (see Figure 67). The assembly is minimal, extremely lightweight and can be removed for the beach or for spectator sports. Design considerations have followed contemporary furniture.

For ease of construction and weight considerations, seating like this would seem ideal for kit-produced cars. But it must be understood that although the aluminum and canvas slings seating is remarkably comfortable *for a 2CV,* in terms of present-day seat design the kind of ergonomically determined seating that shows up in Saabs, Volvos, Chevelles and Pintos is safer and more comfortable. Probably the best-conceived seats ergonomically are in the Porsche.

Most people are afraid that any tampering on their part with the workings of a car will invalidate its warranties, and this fear has kept many gas-saving gadgets or additives from being used. Dr. Robert E. Seivers, Dr. Kent J. Eisentraut and Robert L. Tischer (all chemists at Wright-Paterson) have developed a fuel additive to replace tetraethyl lead as an antiknock component. The compound, cerium 4 (2,2,6,6,-tetramethyl-3,5-heptanedionate), eliminates pollution caused by the burning of lead and gas. Another product is the "MBI Pre-combustion Catylist" made by Mo-Bile Industries of Piscataway, New Jersey. A thick bimetallic screen made of cadmium and nickel is slipped between carburetor and intake manifold. As the fuel passes through the screen, a charge influences the behavior of the fuel/air mixture, increasing fuel efficiency by up to 22 percent.[14] These and other devices could be implemented but are not because of the reluctance by the public to view their cars as anything but a "spectator sport."

But kits need not be tailored only to constructs the size of an automobile. One of our postgraduate students at Manchester Polytechnic was Michael Morris, who was afflicted by a serious stuttering problem. In conversations with him and paramedical personnel it emerged that one way of helping stutterers is to train them with a metronome. Stutterers can learn to pronounce each syllable with each beat of the metronome.

Mike then developed a flashlight-battery-powered, pocketable electronic metronome. The device (of which both pictures and a possible circuit diagram are given in Figures 68, 69 and 70) is smaller than a pack of cigarettes. The small button is an on-off control. The larger turning knob varies the beats per minute so that the stutterer can continuously vary the rhythm to fit his own patterns of syllable formation and speech. A standard earphone plug completes the assembly. The case, in this first

68 Variable-rhythm electronic metronome to aid stutterers, designed by Michael Morris, a post-graduate student at Manchester Polytechnic, in 1974

69 The same device, miniaturized to the size of a fountain pen. We redesigned it to eliminate moving parts; it is light-sensitive and controlled by being "played" like a flute.

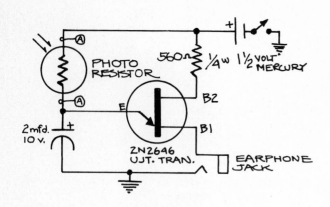

BEATS PER SECOND ARE
ADJUSTED BY COVERING
OR UNCOVERING THE HOLE
IN THE CASE
WITH THE THUMB
IF KNOB CONTROL IS DESIRED
SUBSTITUTE CIRCUIT BELOW
FOR PHOTO RESISTOR.

30K ¼ W.

10 K
LINEAR
POTENTIOMETER

70 Blow-apart drawing and circuit diagram for the device

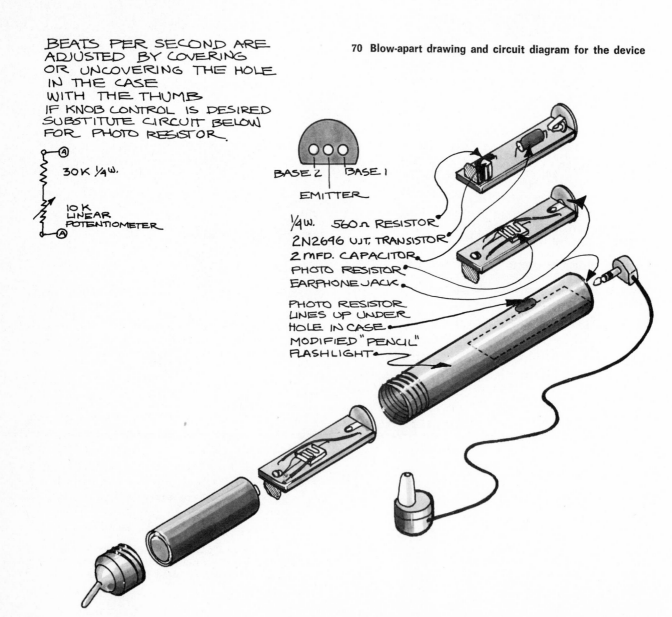

BASE 2   BASE 1

EMITTER

¼ W. 560 Ω RESISTOR
2N2696 UJT. TRANSISTOR
2 MFD. CAPACITOR
PHOTO RESISTOR
EARPHONE JACK

PHOTO RESISTOR
LINES UP UNDER
HOLE IN CASE
MODIFIED "PENCIL"
FLASHLIGHT

prototype, is vacuum-formed plastic. Total weight including battery and earphone is seven ounces.

◐ Other specialized medical equipment and devices for the handicapped could easily be produced in kit form. Especially since the variations even within one category of handicap (such as *spina bifida,* paraplegia, cerebral palsy, etc.) can vary tremendously, the multiple options of kits, described in detail in our examination of the automobile, would force new solutions into being.

◐ For example, another postgraduate student from Manchester, Roger Dalton, has developed a prototype for a kit to build a training treadmill that a blind person can use for walking exercises (see Figure 71). The device has been tested successfully.

◐ But while one can foresee millions of people building consumer products (including vehicles) from kits in the future, there is a secondary and even more socially compelling spin-off. If we assume that kit production will gain ready acceptance by manufacturers and public alike, and that kit producers will switch to large-scale production procedures, this will still leave a huge reservoir of people who, for various reasons, can't or won't build things from kits. Their reasons may include lack of time, unwillingness to spend leisure time at kit-building, lack of tools, lack of mechanical ability or aptitude, visual or motor handicaps, or unwillingness to engage in manual labor. Still, with kits made in large production runs, someone could build the kit for them. A friend, or a local handyman perhaps.

◐ It would make more sense, though, for autonomous local organizations or manufacturers, or a combination of the two, to start up secondary manufacturing facilities. Such a facility would buy kits in quantity at wholesale prices. The products, however, would be put together on an individual basis and with total individual input and feedback from the future user. This would make all possible options available to the end user, thus reversing most of the present-day decision-making powers, which rest with producer rather than purchaser. The assembled kits would sell at retail prices. Since, aside from the kits themselves, there is no raw-material input, and since, aside from assembled kits, there is no output, such secondary manufacturing facilities could be located anywhere. They would need only access to a transportation link such as a canal system, the rails, or truck lines.

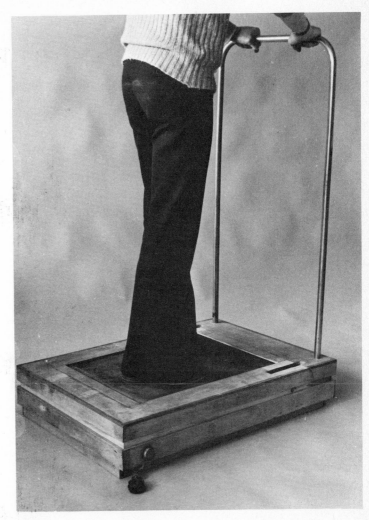

71 Completed do-it-yourself kit for walking treadmill, to be built by the blind for themselves. First working prototype, designed and built by Roger Dalton, a postgraduate student at Manchester Polytechnic.

◐ Again, a facility of this sort could be any size, consisting of as few as two or three people. The process of decentralization could thus begin to serve such differing constituencies as sections of Appalachia without forcing people off their farms and without dislocating present-day traffic networks in the region, which are more than adequate except for big-business purposes. Other constituencies might include dropout communes, so-called slum areas, or suburban towns that offer no employment opportunities to women who cannot commute to work. Other locations might include summer vacation areas that are unsuitable for winter sports—and conversely, skiing resorts that are semideserted during the rest of the year;

penal institutions; shelters and half-way houses for the slightly handicapped; and even mental institutions. Such a system would also create employment opportunities for seasonal workers such as farm laborers, vacationing high-school and college students and many others.

While such a system would be more responsive both to the personal desires of the users of the products as well as to the individual needs of the workers, it would also be much more flexible than present-day manufacturing, assembly, wholesale and retail industries. Because of the wider range of options, innovation could be introduced very rapidly, as could safety improvements. Since secondary manufacturing facilities would be flexible in size, and located virtually anywhere in the country, regional needs and local desires could be easily catered to by assembly teams. The self-confidence and pride in workmanship that has been a traditional American trait would be restored to workers. Engineers, designers and administrators would have thousands upon thousands of feedback loops established to maker and user alike. For the first time the people would control the means of technological production.

5

Would You Like a Fuzzy Photo in Eight Minutes?

In reality the Gross National Product is simply an index of the resources we have destroyed during the years.

—Gordon Rattray Taylor, *How to Avoid the Future*

Frequently things don't work because they were never intended to. A large number of products in our society were brainstormed into being, engineered to meet imaginary performance requirements, styled cosmetically to appeal to acquisitive drives, and then marketed through sales organizations and media until people were persuaded to buy them. When the product fails, as is too often the case, more engineers are brought in to fix it up through additive design. Frequently, trivial artifacts will be produced with minor variations for decades.

Developing the criteria to judge the quality of a product is not that difficult. These would include: materials; workmanship; reliability; safety; durability and life-span; ease of maintenance and repair; accessibility and availability of parts; finishes; and associational and aesthetic values. This is the stance from which such organizations and publications as *Consumer Reports, Consumer Reports Buying Guide, Moneysworth, Consumer Bulletin, Which?* (in England), Nader's Raiders and other consumer buying guides and organizations examine products. But the existence of the product *per se* is never called into question. The question "Should it exist, and if so, why and for whom?" is never asked.

The questions that should be posed are:

1 Do you really need it, or are you being persuaded through advertising and gimmickry that it is something you want?

2 Will something else serve the purpose? Are there substitutes that you already own which will perform the same, or similar, function?

3 Can you share it, rent it, borrow it or lease it? (See Chapter 2.)

4 Can you buy it used? (See Chapter 8.)

5 Can you build it yourself from plans? (See Chapters 3 and 4.)

6 Can you build it from a kit? (See Chapters 3 and 4.)

If your answer is no to these six questions, then go ahead and buy it new!

This hierarchical checklist is quite rigid. We discussed it for a long time, attempting to find an example that we could feed through this sixfold filter. Even in discussing some exotic and recherché examples, we found almost nothing that would pass all six points of our list.

(And some examples can be *quite* exotic. We work as advisers to the World Health Organization in Geneva, Switzerland, and are designing a Tropical Diseases Diagnostic Kit to be used for health services on a village level in Third World countries. The kit will be used by midwives, witch doctors, village headmen, shamans and medicine men. The guts of the kit are some color photos showing typical rashes, skin conditions and eye diseases, but it was found that people in some primitive societies cannot make the perceptional leap to a second-order abstraction. That is, they cannot relate a color photo to a real person and make comparisons. They can, however, successfully compare two color pictures. Hence World Health's anthropologists suggested that an instant-picture color-print camera be included in each kit as a diagnostic tool.)

After much deliberation, discussion and great effort, we have finally isolated one product that might successfully pass the test. It is the new Volvo catalytic converter mentioned in the previous chapter, and offered to the public only during the last few weeks. Obviously, for social and ecological reasons it is something that is needed. Since it is ten times more efficient than any other catalytic converter, nothing else will serve the purpose and, assuming that you need to ride in a car, nothing else will perform the function. It cannot be shared, rented, borrowed or leased. Being brand-new, it does not yet exist in used condition. It cannot be built from plans, and no kit exists from which it can be built. Therefore we say go ahead and buy it (always assuming that it is compatible with other automobiles).

There may be an important corollary next step: If you discover a need that effortlessly moves through all six points, then go ahead, design it, invent it, build it! While the invention of something totally new may at first seem daunting, the fact of the matter is that a technologically mismatched society that begins to listen to new voices from the oppressed, the disadvantaged and the poor will create things with new awareness and also with new sensitivity to real emerging needs.

In recent years, various instant-processing folding cameras have been marketed. The difference between any two cameras may be mainly cosmetic: Whereas the "economy" model's case may be made of a plain plastic, the "top-of-the-line" case may be made of a plastic that gives the appearance of metallic ruggedness and precision. The camera comes with lengthy and endearingly worded instruction manuals, which provide toll-free telephone numbers in case of problems or malfunction. The film packs have warnings and cautions printed all over the package as well as in a separate flyer tucked into the box, with further cautionary phrases stenciled on the pack itself or attached with sticky tabs.

For a long time, only a lucky customer came away with a film pack whose battery hadn't gone dead while sitting on the shop shelf. Since the film-pack developers contain a caustic acid, warnings are attached to both package and product.

The cameras, their parts and the film packs may also have to be protected from temperatures that are the summertime norm in most of the United States; from winter low temperatures that seem mild compared to average temperatures in at least twenty states; from humidity occurring in most of the Midwest and South; and remarkable for what are basically outdoor cameras, they must be protected against prolonged direct sunlight.

Let's examine both extremes. On a very hot day the caustic acid in a film pack may expand and leak, and a built-in battery may blow itself out of its little Mylar bag. The combined acidic goo now oozes undeterred through the camera, eating, destroying or at least discoloring various parts. If left on the back shelf of the car on a hot day, camera and film pack may form a wet puddle of melted material. (Yes, we know the manuals say not to do this, but people are forgetful: It still ends up as a $195 puddle.)

On an extremely cold day, by contrast, when a camera is opened the bellows might shatter into a thousand fragments. The electronic shutter mechanism can freeze tight, thus burning out the little motor. The plastic body may lose its impact resistance. Finally, if it is both cold and wet, the leaves of the shutter could freeze and destroy the small electronic arm linkages.

Regardless of temperature or humidity, the final picture of instant-processing folding models is rarely consistently color-correct and only accidentally, if ever, in *critical* focus. While it takes only 1/125 of a second for an exposure, you wait around for something like eight minutes while the fuzzy picture develops itself—another "spectator sport." Considering that other, cheaper cameras develop pictures in sixty seconds, this seems a seven-minute retrogression.

One major company has in all good faith spent immense sums of money to improve their product and has succeeded in extending the shelf life of the film pack. However, this does not change the fact that the entire concept of folding was an attempt to trade, through heavy advertising, on people's desire for instant ego gratification. As a result, millions of people have spent billions of dollars to add still another camera to their others, quietly obsolescing in closets and camera bags.

Foisting a product that is not needed on the public is bad enough; but when the product is high-energy consuming and/or polluting as well, it creates a problem with deeper societal and environmental ramifications. This may be the place to examine recent media-advertised "innovations" and "improvements" that fall into this category.

The historical change from all-tube television sets to hybrids (combinations of tubes and transistors) to today's total solid-state "instant-on" TV sets is a path strewn with casualties, some of them regrettable. We remember, when early TV sets misbehaved, taking out the tubes and carrying them to the nearest drugstore where a TV-tube-testing machine stood in solid majesty. After subjecting all the tubes to plug-in, plug-out diagnosis, accomplished through the turning of dozens of little knobs, it was a real pleasure to find the one that was bad, buy a replacement, take it home, plug it in and contentedly settle back to watch the Sunday edition of "Omnibus."

Nowadays when your solid-state, instant-on color TV goes on the blink, you phone the TV repairman for an appointment. After you have missed both the winter and summer Olympics, a white-frocked, faintly surgical-looking expert finally arrives for his diagnostic house call (at $40 a throw). Some days later the correct circuit board will be replaced and, to the accompaniment of a bill for $119.82, the set is operable again.

But the inherent problem with today's solid-state TV set is not with repair costs but with energy wastage and potential danger. Solid-state television receivers are advertised as being totally transistorized. This is true only up to a point. The biggest tube of all—the picture tube—still remains. Like all other tubes, it requires two voltage supplies: a high-voltage supply that reacts instantaneously, and a low-voltage filament supply that is used to provide power to a glowing element. In the past this element required a warm-up time of one to two minutes before the image appeared. Trading on the impatience of people who want the image on in the very microsecond in which the switch is closed, the television industry has more recently developed the "instant-on" set. This requires a continuous low-voltage trickle to the picture tube, which remains on as long as the set is plugged in, whether or not the set is turned "on." This little

tidbit is usually not communicated to the public. One drawback is that there is a constant energy drain. Another is that, with the set on all the time, a circuit shorting out poses the hazard of TV fires or electrical-wiring fires in the wall.

There is no way to contend with the "instant-on" problem, since nearly all newer sets have been "improved" in this manner. As a minimal safety precaution we suggest that you unplug the set when away from home.

This is not the place for a pop-sociological discussion of the impact of TV commercials. Instead we provide a photograph and circuit diagram for making a simple "commercial-killer." (See Figures 72 and 73.) When a flashlight is beamed at the photocell above the TV set as the commercial comes on, the sound is killed until a second flash of light restores it when the program resumes. It's a great treat to be able to watch a TV feature film or documentary without the constant interruption of mealy-voiced hucksters pushing deodorants, instant soups, drive-in churches or breakfast goodies. (While the unit does work well, we apologize for the circuit design. It is a concoction that could probably be improved and simplified by any competent electronics engineer, who might also add a thirty- and sixty-second timer, that being the average length of commercials.)

Self-cleaning ovens in electric ranges are another consumer "improvement." An electric range, even in its standard version, is one of the highest energy consumers in the household. When a self-cleaning oven is added, energy consumption climbs rapidly. But, as with the color TV set, this is only where the story begins. The self-cleaning cycle raises the interior temperature of the oven and "removes" spilled food by reducing it to gas and water vapor at a temperature as high as 1000 degrees F. (This is nearly half the maximum temperature reachable in a ceramic kiln, or a crucible in which gold is melted.) Let us repeat: *No* food residue is *removed*. It is turned into water vapor that reduces down to an ash that must be wiped out, and gas.

The process takes two to four hours. Since it usually takes place in the evening or during the night, it is unsupervised. But there isn't much to watch anyway; you might get a greater kick out of watching the dials on your electric meter spin. According to an obviously self-serving maintenance

72 "Commercial killer" atop TV set. This is activated with a standard flashlight.

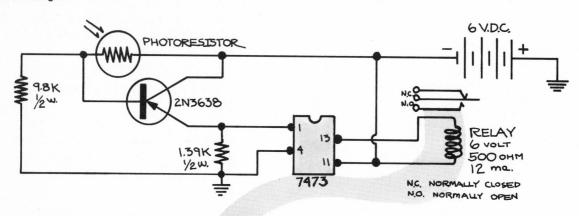

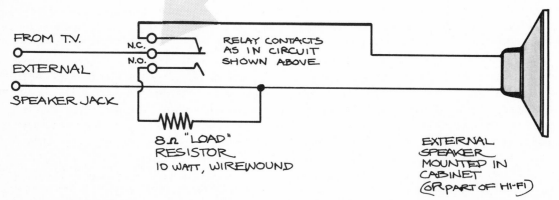

73 Circuit diagram for this device

manual: "The entire operation . . . costs less than the can of oven cleaner needed for a hand-cleaning job." Current prices place a small can of spray-on oven cleaner at $1.89. Comparing the cost of "pyrolytic" oven cleaning to the cost of a chemical cleaner means comparing two unnecessary products. And there are the further drawbacks that any charred substance that survives the holocaust and remains firmly burnt onto the interior cannot be scoured away while oven cleaners and abrasives damage the special surface of self-cleaners and destroy their self-cleaning properties.

☯ Baking soda and water, or hot, soapy water will do the same job at a tiny fraction of the cost, although the savings are small considering the labor and time involved.

The extreme heat used in the self-cleaning process requires heavy insulation, a thick oven door, a fiber-glass door seal, finishes designed not to crack at high temperatures and many safety devices. Three or more controls must be engaged and the oven door has to be locked. As the temperature moves above 600 degrees, an automatic closure freezes the door lock until the cycle is over.

Since these ovens are made and inspected under the same mass-production conditions discussed in the previous chapter, units that are faulty and far from fail-safe may be installed. Conceivably, with tougher business competition in the present depressed market, cost cutting and so-called value-engineering procedures may also result in units that are less than safe. The possibility for such units setting surrounding cabinets, the kitchen or an entire building ablaze should at least be carefully considered. The other fact to be considered is the high initial cost of the unit.

☯ Gas stoves, by contrast, now come with continuously cleaning ovens that require no extra adjustments, extra energy or high heat.

Refrigerators exist in a bewildering variety of types: manual defrost, self-defrost, permanent frostless as well as refrigerator-freezer combinations including "frostfree" features, and with other possible options such as ice-makers and an exterior-mounted "convenience center" that dispenses chilled or warmed drinks and ice cubes.

First, it should be said that most refrigerators are needlessly large in size. Their very size promotes the chilling of many foods that need not be in a refrigerator at all. Basically flavorless and tasteless fruit, hybridized for good cosmetic appearance and long shelf life, is made palatable by chilling it for days or weeks. That peculiar replica of polyurethane foam, sold as white bread, is also made slightly less unpleasant by being stored in the "fridge" rather than in a breadbox.

Secondly, refrigerators are underinsulated and slightly underpowered.

☯ With marginally better compressors and heavier insulation, the *average* life-span energy bill (now running about $800) could be reduced to less than $300 (based on a ten-year life).[1] Energy costs for an eighteen-cubic-foot "frostless" refrigerator are *three times as much* as for a twelve-cubic-foot manual-defrost model.[2]

Even this comparison is misleading as there is a further energy saving in manual-defrost-type units, since the entire refrigerator is turned off while being defrosted. Since, on the average, a refrigerator is defrosted about four to eight times a year, the energy saved is worth the amount of work involved.

The ice-maker feature adds a further, incredibly large energy drain. Moreover, few people realize that instead of plugging in, as other refrigerators do, it is necessary to plumb in a thin copper water line to supply the ice-maker, in addition to the usual electrical requirements. In difficult cases this installation alone could cost nearly as much as the refrigerator.

☯ The refrigerator is an absurdly simple electrical appliance that has undergone no significant changes, aside from cosmetic styling and the addition of useless—but energy-wasting and expensive—gimmickry. Hence refrigerators can be bought used. (For more information on this, see Chapter 8.)

A recent market study of appliances in Great Britain devotes an entire chapter to dishwashers under the heading "The Appliance Nobody Wanted." We quote from the introduction: "It seems that the British like washing up, or perhaps they do not dislike it enough to invest in a machine to wash dishes for them."[3] Sales have been stuck for years at 30,000 units per year in Britain, while 500,000 units per year are sold in West Germany, and even in an economically depressed country like Italy, sales have reached 250,000 per year.

British objections to the dishwasher usually are as follows: Dishwashers are too expensive. There is no room for them in the average kitchen. They will not wash everything. Most of them do not work properly. They damage delicate china, break glasses and scratch up most cutlery. Furthermore, for best results, dishes have to be scraped and pre-washed before being put in the dishwasher. Usually two or more chemical cleansers are needed: a special dishwashing powder, as well as a rinsing agent to keep glassware from "spotting." Dishwashers are noisy, waste water in the input stage, and pollute water thermally and chemically in the drain cycle.

Manufacturers' counterclaims to this last argument usually explain that "hand-washing" and "hand-rinsing" of dishes use "sixty-three times as much water as your new dishwasher." This phony statistic is based on washing the entire load of dishes under running water and rinsing them in the same way. It is further distorted by the fact that it includes the washing and rinsing of pots, frying pans, coffee percolators and other items that would never be washed in a dishwasher at all. (Unless one of the newly touted "pot-scrubber" dishwashers has been purchased.)

Basically dishwashers all have the same simple method of operation: After they are loaded and a timer-cycle activator is set, they go through a hot-water (140 degrees) wash cycle with detergents added, a rinse cycle, and finally a drying cycle with forced air added in newer models.

This basically simple procedure has been pseudotechnocratically improved, so that now one of their "features" may be listed as "Ten Push-Button Selections for Automatic Dishwashing." This could be a quote from many a manufacturer's top-of-the-line model; their economy model may have only two push buttons. Note that dishes and flatware for both models come out exactly the same.

The most ironic twist is the attempt to cash in on recent concerns about energy waste and cost. Manufacturers who, not too long ago, had added as an innovation a heated air-drying cycle that consumes more energy than the ordinary air-drying process now provide the user with an option to cancel out the heated air-drying feature, as explained in the following advertisement: "POWER MISER CONTROL . . . set Power Miser control on 'cool dry' and the 750-watt heating unit shuts off during the drying phase. You save electrical energy, as cool air dries your kitchenware."[4]

The logical outcome of adding a button to eliminate a previous one might be a machine with forty push buttons, twenty of which cancel the work of the other twenty—the machine finally doing nothing at all. Thoughtless additive design has given us first a hot-air drying cycle, then a fan-forced system to shorten drying time, and, finally, a control to eliminate the first of these three. The question is, was the heating element ever really needed in the first place?

As mentioned earlier, dishes must be scraped and rinsed before being put in the dishwasher; delicate glassware and some pans and small appliances must be hand-washed and hand-rinsed. With a little more effort the dishes, too, could be hand-washed and hand-rinsed, eliminating the need for a dishwasher entirely. After hand-washed dishes come out of the hot rinsing water, they can be air dried in preference to the unsanitary method of cloth drying. Housewives, however, object to this from an esthetic standpoint: They don't like a rack full of drying dishes sitting on their counter for all to see.

There is a little-known solution that has worked well in Finnish kitchens for the last fifty-five years. A standard dish cabinet with an open bottom is built directly over the double sink. The cabinet has sliding dish racks as drawers—similar to the drying racks made of wire baskets dipped in vinyl now made by Rubbermaid and other companies. Dishes, glasses and flatware are placed in these racks after being washed and rinsed. The racks are then slid back into the cabinet and the doors closed. The cabinet is lit by incandescent tube lights that help to dry the dishes, as well as provide lighting over the sink. Any excess moisture drips down through the open cabinet bottom into the sink.

It must be emphasized that these dish racks organize dishes, glasses and flatware and are also the only permanent storage place for dishes in a Finnish house. In Figure 74 we show a drawing of one of these cabinets for do-it-yourselfers, who could easily build this or an improved version for their own needs.

We could go through a whole list of appliances and gadgets like electric carving knives, electric toothbrushes, toasters, snowmobiles, disposals, TV-simulated games, facial saunas, orbital and re-

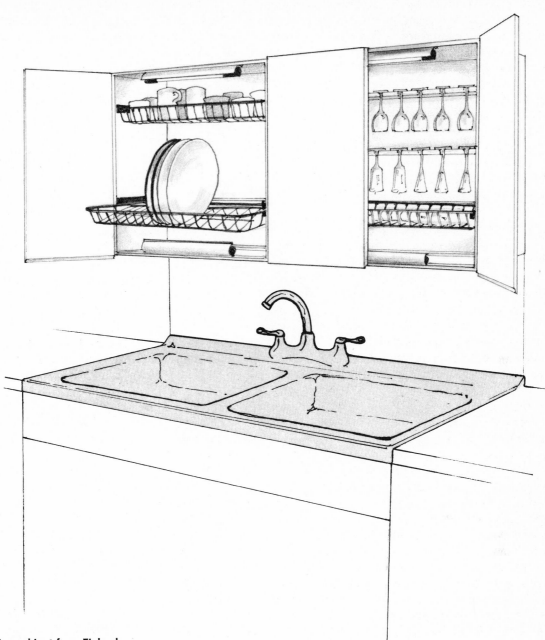

74 Dish-drying cabinet from Finland

ciprocal massage instruments, and much else merely to repeat the points already made. Instead we would like to use another case history, that of the can opener.

In the majority of can openers the lip of the can is inserted between a drive wheel and a cutting blade. The can then rotates so that the entire rim passes through this assembly, separating the lid, which supposedly will be held by a magnetic arm. The can is now opened and the lid thrown away.

In the real world things don't move that smoothly. Some cans are extremely difficult to position in the opener, some impossible. Once engaged, the drive wheel can frequently slip (especially when worn), resulting in an unopened or partially opened can.

When the cutting action does begin, several other things happen, too. Some of the contents of the can slop onto the cutter and wash some of the caked residue from previous cans into the can

75 "Sesame" can opener by Westinghouse

76 Comparison

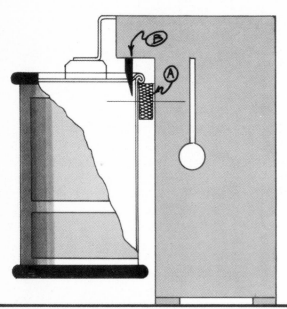

CONVENTIONAL OPENER

KNURLED ROLLER Ⓐ DRIVES
CAN AROUND WHILE CUTTER
Ⓑ SLICES LID AT THE TOP

1. LID IS SMALLER THAN CAN
AND CAN FALL IN.

2. METAL FILINGS CAN CONTAMINATE
CAN CONTENTS.

3. CONTENTS TEND TO COLLECT
ON CUTTER AND CREATE
RUST.

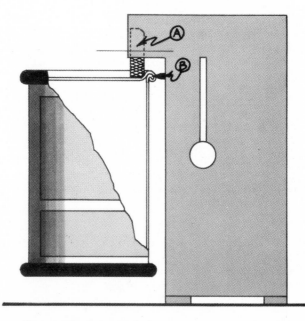

"SESAME" OPENER

KNURLED ROLLER Ⓐ DRIVES
CAN AROUND WHILE CUTTER
Ⓑ CUTS FIRST LAYER OF
ROLL-OVER AT SIDE.

1. LID IS NOW BIGGER
THAN CAN AND CANNOT
FALL IN - CAN BE USED
TO RE-SEAL CAN.

2. METAL FILINGS FALL
OUTSIDE OF CAN.

3. CONTENTS & CUTTER
DO NOT COME IN
CONTACT.

being opened, thus adding contaminated particles to the contents. (Most people never or rarely clean the cutter blade.) The corrosion of the cutter due to food particles, rust and the liquid contents of cans dulls the blade so that the opener does not operate properly. But even when it does, the shearing action drops small metal burrs into your tomato soup.

In addition, the magnetic arm, which supposedly holds the lid being sheared out of the can, frequently doesn't work, and never works on cans with aluminum tops. The cut disk, being smaller than the can opening, then drops in, forcing you to fish for it. If its razor-sharp edges don't cut you, they may cut the garbage collector.

Finally, a child playing with the can opener might easily get a finger caught between cutter and drive wheel. In fact, since your own view of the cutter is blocked by the magnetic arm, there is at least a fighting chance that you may get your finger(s) caught as well.

By contrast, the Westinghouse "Sesame" can opener does a superior job. Researching how a tin can works, a synectics team utilized the fact that the lid rim is rolled over the can body. While the "Sesame" also employs drive wheel and cutter (now in the shape of a wheel), they are arranged at a right angle, compared to the conventional opener described above. (See Figures 75 and 76.) There is no magnetic arm to hold the lid, because the lid, after being cut, ends up the same size as the can and will not fall in. The cutting wheel cuts through the outer layer of the roll-over so that there is no residue; tiny metal fragments fall *outside* the can. Simultaneously the top edge of the can and the bottom edge of the newly made lid are crimped over, doing away with all sharp edges. The newly created lid can now serve as a closure for the can's contents in the refrigerator. Since there is no magnetic arm assembly, the entire cutting process is visible to the operator, eliminating the possibility of injury.

Westinghouse marketed this device, which was manufactured for them in Japan, in the late sixties. With the phasing out of their consumer-products division, the logically designed and highly innovative can opener also disappeared. It is no longer manufactured.

While the "Sesame" can no longer be bought, vastly inferior electric openers (lately "improved"

77 The "G.I. Friend" can opener

with electric knife sharpeners and even "task lighting") are still available everywhere. There are hundreds of these, basically identical, each one constituting an energy drain both in their manufacture and use, and made by scores of competing manufacturers. The question is, who needs an electric can opener at all? And this is not said in defense of finger-amputating and lip-cutting zip-top beverage cans.

The humble "GI's friend" (shown in Figure 77) and other manual openers can work in all situations in a fail-safe way, possibly excluding only the kitchen needs of quick-food restaurant chains. While not all manual openers work well, the GI's friend, measuring two inches by half an inch by an eighth inch when folded, operates, when unfolded, without any moving parts. It is so simple that nothing can go wrong with it, yet it does a tremendous job of cutting. Some have been in use for forty years.

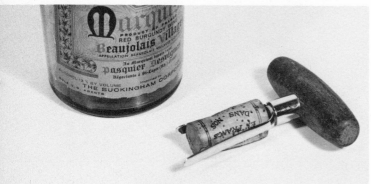

Cork extractor. 78, 79

Waiter's corkscrew. 80,

We see a continuous replacement of good and usually simple ideas by vastly inferior and complicated ones. The corkscrew is a case in point.

Corks in wine bottles are forced in under great pressure, dry out and get crumbly if the bottles have been mistakenly stored in an upright position, and frequently develop a thin film of sugar residue effectively cementing the outer edges of the cork to the inner bottle neck. We feel that ideally no cork remover should penetrate the cork at all, since small fragments of cork will unavoidably fall into the bottle.

⊙ There is a wine-bottle opener which employs two thin metal "fingers" that are forced down between the cork and the neck of the bottle. By rotating it and pulling upward, sticky sugar residues are cut and the undamaged cork is eased out. (See Figures 78 and 79.) It is absurdly inexpensive and, probably for that reason, difficult to find.

⊙ Penetrating the cork, the simplest way is the use of a wine steward's pocket opener (Figures 80 and 81), which employs a lever action to pull a clean cork with great ease. Widely used on the Continent and in restaurants, sturdily built to last a lifetime, it is quite inexpensive but only rarely available in gourmet shops. Other corkscrews employing double levers, or double-threaded ones, also work with a minimum of effort, but again are not too easily found.

Recently the "improvement brigade" has come up with a series of bottle openers in which a needle penetrates the cork. Then either air is pumped into the top part of the bottle, or else a CO_2 cartridge is used to force the cork out. All of these devices carry warnings and can be used only on round bottles. If there are any flaws in the glass, the chances are that you will rupture the bottle and blow glass shrapnel into your groin.

81

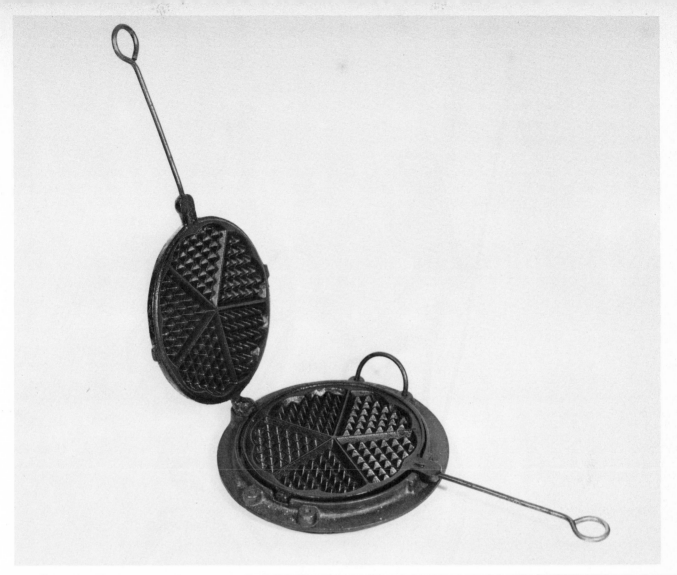

The "flipper" waffle iron
from Sweden. 82,

83

84   A modest device to maim, cripple, or kill a child

The Swedish *"Flipper"* waffle iron is another example of a simple implement that has stood the test of time. It is a cast-iron, hinged, two-sided mold with two handles which is used in conjunction with your regular kitchen range. It is nonelectric, has, aside from the hinge, no moving parts and makes excellent waffles. (See Figures 82 and 83.) It has been in continual use in Sweden since the 1600s and uses no extra energy nor is it prone to damage, malfunction or short-circuits, as are expensive electric waffle irons. A more recent model is Teflon-coated.

It is also becoming difficult to purchase a child's bicycle that has not been ergonomically, disastrously redesigned, forcing the child into bad posture and use of the wrong muscles. In addition, these bikes carry dangerous protruberances. High-rise handlebars are probably the most dangerous, succeeded by "banana seats" (for more than one rider). (See Figure 84.) "Moto-cross-style" bikes and similar models pay little regard to the task of pedaling in an attempt to make the bike look like an engine-driven trail bike. To this can be added the confusion of operator controls; badly built multi-speed gearing that often slips, leading to injuries; senseless clip-on AM-FM radios; and detachable walkie-talkies. Often a bicycle that arrives boxed and only partially joined is difficult to assemble, leading to dangerous riding conditions. The crowning absurdity is clip-on "zoom-varoom" noisemakers that provide the bicycle (at some expense of pedaling energy) with the sounds of a motorized bike (to the annoyance of all concerned).

A list of simple things that work well, but have been displaced in the market by trivia that work badly, could be endless. However, we want to end this section by mentioning the electronic golf-ball set that consists of a "base station" with tiny transmitters located in each of the balls. Four golf-balls with two bleeper-receivers sold for a cool $60 three years ago. For an extra $6.95 you can buy a mounting bracket, which mates the receiver to your $1395 motorized golf cart.

Folk mythology of our technological age abounds with stories about how good ideas have been kept from the public by manufacturing interests. Most of us have heard the story of the everlasting match, supposedly suppressed by the Swedes; and we have all heard about the long-life light bulb that no manufacturer is willing to create. (Except for the Crompton-Parkinson 2000-hour bulb in England, of course. There are also a host of stories, many of them true, about light bulbs that unaccountably continue burning. In Graughton-on-Lyme, also in England, a light bulb in the railroad station merrily illuminates the platform and is doing well in its fifty-seventh year.)

All too frequently, however, the middleman or retailer also suppresses a good idea because he feels that articles that work or wear well will diminish his continuing sales. There is a true story about a nearly everlasting sock that lends support to the belief (as expressed by George Bernard Shaw in his preface to *The Apple Cart*) that durability is unprofitable.

The firm of Simpson, Wright and Low of Sutton-in-Ashfield, of Nottinghamshire in England, manufactured socks under the brand "Essinay" beginning in 1952. They were made from spun nylon—that is, nylon staple, chopped into short lengths and spun like wool. When the socks were new, their appearance and feel were nearly identical to wool socks, but they did not wear into holes. Without any darning they lasted for ten years or more, depending on amount of use and the wearer's habits. Their great warmth and absorbency resulted from the re-spinning process. They "pilled" (surface fibers rolled into balls), discolored after five or six years, because of the felting action, even shrank a tiny bit (less than half a size in five years), but they did not wear out.

After the socks had been discussed by consumer magazines in Great Britain, it was discovered that the makers had stopped production, as retailers refused to sell them. In spite of their many excellent qualities and the low price of about 19¢ a pair, the sock is no longer made because of low merchandising profits.[5] A similar story about everlasting, no-run panty hose is told in the United States.

Despite our pessimistic tone, there is hope. The American consumer is beginning to lose the lust for material possessions and is instead turning to travel, sports and entertainment to improve the quality of life. Pollster Louis Harris says, "The 3-car and 3-bathroom syndrome is dying, if not already dead," and he cites the "quite radical findings" of some of his recent opinion surveys, concluding: "This has tremendous implications for our economic planners. What is important to people is no longer going to be material acquisitions."[6]

Patrick H. Caddell, the president of a Cambridge, Massachusetts, polling firm, says his polls lend support to Harris's statements. He says that the number of two-car owners has dropped 3 percent during the past year and 45 percent of all adults say they will purchase cars much less frequently in the future. Upper-income buyers who are better educated, Caddell says, are most inclined to buy smaller cars and smaller foreign cars. He concludes: "If such changes continue we may well find some alternatives to the importance or status assigned to large material goods and, in the end, important and significant changes in consumer spending habits."[7]

Louis Harris estimates that two-thirds of the American people are more concerned now about quality of life than quantity of goods: "I am prepared to say that in the next five years we are going to see a flattening out of absolute demands for the vast majority of physical goods that we have come to depend on for our economic livelihood. This includes washing machines, cars, TV sets, appliances of all kinds. . . ."[8]

Historical footnote: Since the above was written, there seemed to be a contrary trend emerging with people once again interested in larger cars. More recent reports show that GM's gamble with "down-sized" automobiles is paying off handsomely. According to a recent article, the biggest problem seems to be producing enough of the scaled-down cars. During a time period in which standard-size cars by Ford have shrunk by 20.4 percent and Chrysler has declined 14.8 percent, GM has increased its smaller car sales by 13.4 percent. GM dealers report that motorists "like the improved mileage and the boxy European styling."[9]

We are, apparently, no longer so eager to buy —which is all to the good, because having dealt with much that is phony and false in this chapter, we are about to explore even phonier and more dangerous products in the following one. To help you become even more discriminating, we offer some criteria you can use to determine whether a product is too gimmicky:

1 *Look at the product, inspect each feature separately.*

2 *Does each feature have a purpose and need?*

Electric peanut-butter machines (runaway best sellers during the 1975 Christmas shopping spree), automatic apple corers and similar devices are silly, although frequently their essential bankruptcy is disguised by topnotch industrial design. Obviously you can easily core apples by hand; peanut butter is readily available (including "natural" and organic brands, as well as those made while you wait in specialty shops). If you still insist on making your own, nonelectric machines will do a good job, use no energy except your own muscles (which is good exercise), will not break down and are easier to clean.

3 *Do you have a need for all the features?*

Electric sewing machines come in a broad range of options. Many women buy the top-of-the-line model with automatic snap-in monogrammers, scores of interchangeable drop-in cams for decorative stitches, solid-state foot controls and dozens of other accessories and devices, most of which will never be used, or will be misused, damaging the machine.

4 *Does each feature work, or is it just a gimmick?*

Many box fans are advertised as having "three-speed controls." The user often finds out too late that the three speeds are "High," "Low" and "Off." The fact—cited in an earlier book—that a carpet cleaner now comes with a racing stripe in no way adds to its use.[10]

5 *Is the product too dependent on a "system"?*

The introduction of super-8mm motion-picture film made millions of cameras and projectors instantly obsolete. Conversely the Fujica single-8 movie camera is still made, but is dependent on its own film format, which must be bought in special cassettes and can be shown, with the least trouble, only through Fujica projectors.

The same thing has happened with the introduction of Kodak Pocket Instamatic cameras. Since the negatives for color prints are so minute, the photographer depends on computerized processing in "clean rooms," as otherwise the tiniest speck, enlarged to wallet size, seriously mars the image. When these same cameras are used for minislides, the optical loss in so small an image slowly but irreversibly damages the spector's ability to discriminate. (Another drive to lowered quality, for the sake of "pocketable convenience," comparable to the loss of image quality when 2¼-by-2¼-inch transparencies gave way to the 35mm format.) Any "cartridge" system, such as audio cassettes, eight-track players, video cassettes, film cartridges, dictating-machine cartridges, typewriters using cartridge ribbons, and even ballpoint pen refills and razor-blade cartridges, needs the awareness of the buyer before making the purchase.

By using this fivefold system of questions, you may avoid needless purchases, frequent frustrations, and begin the cure from product addiction.

6

No Roast Tonight—the Lights on My Carving Knife Need Realignment

All beauty that has not a foundation in use, soon grows distasteful, and needs continual replacement with something new.

That which has in itself the highest use, possesses the greatest beauty.

—From the *Shaker Millennial Laws,* 1795

Most things we buy come in packages. In this chapter we try to explain why packages don't work, at least for you, the purchaser. We also try to explain why often the contents of the package don't work either. Our reason for combining these two points in one chapter is simple: It is getting harder and harder to tell where the package leaves off and the product begins. Or vice versa.

"Consumer psychology," that frenetic, zany field that gave us "Pet Rocks" (a few small rounded pebbles from a Mexican beach, beautifully packaged, with an amusingly written manual on how to feed, train and keep the rocks, which became, at a cool five bucks a throw, a Christmas gift sensation all over the country), has also given us "canned air." This is, of course, another gag gift. You buy a can, identical to a zip-top beer can, except that the label announces it to contain "Colorado Mountain Air" or "Fjord Breezes from Norway." As we said, it's just a gag.

In fact, air has not only been canned, but compressed as well, and is now successfully sold under the trade name "Dust-Off" to photographers everywhere. The idea is to blow lint off the front half of one of the nine lenses that make up a modern telephoto lens. Dust-Off comes in a studio size, an economy size and an "emergency" gadget-bag miniature. Question: Which is the product and which is the package in this case?

Of course, the pressurized air comes in an aerosol can. That means that the package itself is expensive, hazardous to the individual user, and may be potentially dangerous to all life on earth. In June 1975, Oregon became the first state to ban the sale of aerosol cans using chlorofluorocarbon compounds as propellants. These propellants are used mainly for hair sprays, insecticides, deodorants and oven cleaners. A government report, based on the studies of a scientific panel from the National Academy of Sciences, considered the release of chlorofluorocarbons into the atmosphere "a legitimate cause for concern." These propellants deplete the ozone layer that shields the earth from the most intense of the sun's ultraviolet rays. This is suspected of causing an increase in the risk of skin cancer at a rate of 2 percent per year for each 1-percent decrease in atmospheric ozone.

F. Sherwood Rowland, of the University of California at Irvine, explains that the change begins when sunlight breaks up chlorofluorocarbons, releasing atoms of chlorine. These atoms react with the ozone—a form of oxygen whose molecules contain three atoms instead of the normal two—and convert it to ordinary oxygen, which is rather ineffective as a shield against ultraviolet rays. Were the use of these propellants to continue at the present rate of about two billion pounds a year, it could cause a 13-percent depletion in the ozone layer and lead to possibly 80,000 extra cases of skin cancer a year, as well as other side effects.[1]

We have already argued in an earlier book that aerosol cans are hazardous in other ways as well.[2] For one thing, they tend to explode if exposed to extreme heat or intense cold. There are several cases where children have thrown aerosol cans into a barbecue fire and were injured by exploding shrapnel.

Aerosols also may fail to work properly. Due to manufacturing difficulties, aerosol products frequently leave the factory with full contents but without a propellant charge. Nothing can be more frustrating (or wasteful) than a full can of paint or underarm deodorant that refuses to work and that cannot be salvaged in any way. Also frequently the valves get clogged after a very short time, resulting again in frustration and waste.

The final disadvantage of aerosols is that you get less content for more money; under all circumstances, aerosol cans can never be completely emptied.

◐ If you still want to get that lint off your lens, several German and Japanese firms make a small rubber bulb to be squeezed by thumb and index finger. The bulb terminates in a camel-hair brush. The whole assembly costs less than a medium-size can of Dust-Off and will, with careful use, last long enough so that the photographer can bequeath it to his children.

Manufacturers tend to escalate their offerings. Shaving foam almost completely supplanted brush and shaving soap among most men. When that market was fully saturated, a new beachhead had to be established. The most recent offering is "hot foam." This crucially needed breakthrough has been achieved in three separate ways. First and simplest, the can may just be held under flowing hot water. (This, in addition to all the other problems of aerosols, also wastes heat and water.) The second relies on an electrical plug-in unit, into which each can is inserted to heat up. The third is based on thermochemical reaction that takes place as the foam leaves the can and comes in contact with oxygen.

In each case, the best of the space industry and the technology that sent man to the moon have made it possible for you to expend money and energy just to be able to push a button that provides a tepid puddle in your hand.

Why this frantic surge for "improvements" in shaving? If you listen to your teen-age son you might hear: *"No father taught me how. I learned shaving on Channel Three."* The "rules" for shaving, like hair care for women, are largely acquired from TV commercials. Since shaving is not a pleasant activity, industry continuously tries to come up with new products that promise to make shaving a sheer delight. The rule is simple: They (consumers, that is) will try *anything;* even if the sucker doesn't buy it, Santa Claus is sure to stuff $29.95 worth of thermoelectric activator, plus three canisters of foam, into the stretch-nylon Christmas stocking.

The dislike for shaving has also given us a bewildering array of razors, razor handles, single and double blades and electric razors. Since about 40 percent of all men find electric shaving painful, many men own $50 or $80 worth of unused electric razors, which they have buried in some drawer.

Then there are both electric and battery-powered toothbrushes. What goes on the toothbrush is still toothpaste. Some of it is a gel, and now also comes in an aerosol can. But what's wrong with tubes? Tubes are in fact a superb minimal package. In Europe not only toothpaste but mayonnaise, mustard, anchovy paste, marmalade, jams and even caviare come in tubes.

◐ We suggest tubes as an excellent alternative to aerosols. But even tubes (or rather how they work and how we empty them) can be improved.

◐ In northern Europe, tubes of edibles, toothpastes and glue come with a small "tube-squeezer," which we illustrate in Figure 85. It is a giveaway made of wood or plastic. You insert the bottom of the tube in it and then just "crank" the contents of the tube out as needed. When the tube is completely empty, it can even be unwound, the metal recycled and the tube-squeezer used again on countless other tubes. Since the only moving part is your wrist, this is another item that you can will to your children.

◐ A Swedish manufacturer makes another device that completely empties tubes. As you can see in Figure 86, it adheres to the bathroom wall. The tube is mounted in it and a plunger dispenses the toothpaste directly on the brush. The cap is never needed again. Through the use of a vacuum plunger this device totally empties the tube, thus no contents are wasted.

85 Tube squeezer

86 Vacuum-tube emptier

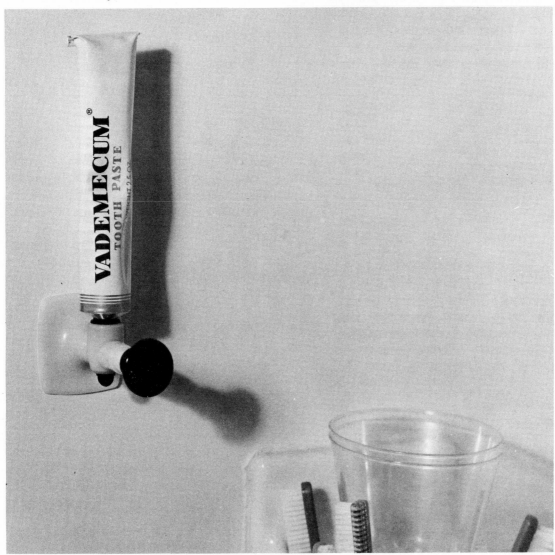

Before we leave the delightful world of aerosol cans, one more item needs to be added. At least one company offers a drain cleaner in an aerosol can. This satanic contraption is inserted, nozzle-down, in the clogged-up drain. By applying downward pressure, the contents are blown into the sink under great pressure, hopefully clearing the obstacles. If you have a double sink and neglect to seal the other drain, chances are excellent for a quick shower and finding the obstruction embedded in the ceiling.

A few times, descriptive terms are endearing. Possibly the most honestly marketed device is an aerosol bomb. An insecticide and fumigant, it is placed on the kitchen floor, the nozzle is pried open and then, according to the manufacturers' gentle caution, it is thought best that you abandon your house for three hours. Children or household pets may not be aware of this state of self-inflicted siege. . . .

◓ Most recently, in part because of Oregon's new law against aerosols, as well as a fear of being legislated out of existence in other states,

88 Simple "blow" sprayer

87 Pump-type sprayer

manufacturers have begun introducing hair sprays, deodorants and other liquid sprays with a plunger-type dispenser. This is usually made of plastic and is manually pumped (see Figure 87). It is certainly a better way of dispensing liquids. Some products have used this type of dispenser for years—Windex, for example, and certain types of hand lotions—so it is somewhat upsetting to see the snake-oil brigade from Madison Avenue pushing this on TV as the newest, latest technological breakthrough.

◓ There are other ways of getting liquids to dispense easily, or turn into sprays. Atomizers have crowned perfume bottles for over a century now and work exceedingly well by breaking liquids up into a small droplet mist.

◓ Art students, until recently, were trained to blow a fine spray of fixative as a protective layer over pastel drawings. The device involved (see Figure 88) consists of two tin tubes a little smaller than drinking straws and set at right angles. They can still be bought at hobby shops and are now used for applying a fine water or lacquer mist to the tissue wings of model airplanes.

◓ Squeeze bottles lend themselves to either rapid or slow dispensing. There seems no practical reason why their use should be largely confined to antihistamine nasal sprays and Elmer's Glue.

◓ Portioning out tiny measured amounts can be done with a syringe, as known to everyone with a sewing machine, or someone who has lubricated precision parts.

◓ For large amounts of liquids, a spigot on a barrel serves well. In the South, for example, apple cider still comes in small casks that fit the refrigerator shelf. In the darkroom most chemicals are stored and dispensed in much the same way.

Some things are overpackaged unnecessarily. Milk, for instance, comes mostly in waxed cardboard containers or their plastic equivalents; bottles have been phased out nearly everywhere in this country because of handling costs, sterilization expenses and the alleged detrimental effect on milk of sunlight through a clear bottle.

⊙ Here again there are alternatives. In many parts of Europe, milk comes in a dark-brown glass bottle, thus screening out the notorious "killer rays" of the sun.

⊙ In Canada, half-gallons of milk are delivered to the home in completely limp, unstructured, thin-gauge plastic "pillows." You take the totally floppy milk pillow and insert it in a plastic or glass holder that has a handle. The packaged milk adheres to the interior of the holder through gravity; the holder in turn provides strength. In order to pour, one small corner of the milk bag is snipped off with a pair of scissors. (See Figures 89 through 92.) The bags are disposable, and a biodegradable version is now being tested.

Going to the drugstore and buying, say, deodorant, razor blades, lip balm and a few other necessities for a trip, one emerges with a large shopping bag full to the bursting point. On reaching home and unpacking the purchases, try the experiment of putting the plastic blister packaging and the cardboard backing boards (along with some broken fingernails) back into the shopping bag. You will find that the bag seems as full as before, even though your purchases fit snugly into a toiletry case.

The reason for large backing boards, accompanied by plastic blisters, is to cut down on pilferage. At the same time, one can crowd one side of the backboard with advertisements, giveaway offers and misleading information. The back is printed with diagrams and an involved set of opening instructions that would have raised an appreciative snigger from Torquemada. Small, comparatively expensive cosmetic items such as eyelashes, eye liner, eye shadow, eye hi-liter, mascara, "blush," and plastic fingernails (including adhesives for them) are commonly packed this way, to name but a few. Batteries also are blister-packaged now, which makes it impossible to test them and increases the chance of getting a half-dead or dead one. We just bought two very exotic alkaline "button" cell-batteries, blister-packaged, at $4.50 each, and discovered to our dismay that they were both dead. The store refused to make good on the basis that "if it's blister-packaged it's not our responsibility!"

In general, blister packages prevent the customer from ascertaining workmanship, quality and shelf exposure (age) of the product.

Men's shirts are often lovingly enshrined and overpackaged. They are stiffened with a cardboard insert (lately plastic), which is usually covered with tissue paper. Around the collar there is an inner sleeve, of plastic or paper, and a triangular, heavy-gauge clear plastic sheet that slips under the collar button. The shirt itself is held together with six to twelve straight pins. One of these you invariably neglect to remove, and it becomes hopelessly entangled with trousers or legs when you attempt to wear the shirt for the first time. If the shirt has French cuffs, additional plastic links are inserted to secure them. This whole crazy assemblage of plastic, cardboard, metal pins and tissue paper, together with the shirt, is in a plastic sack that in turn sits in a box. Little tags with washing instructions, and extolling the virtues of the shirtmaker, are attached to one or more buttons with string.

Since obviously someone must pay for all of this hardware, as well as for the labor that goes into inserting it, and since neither shirtmaker nor retailer bears the ultimate cost, you can see where overpackaging becomes directly costly to you. (It must be added that new shirts are commonly so overstarched that they creak.)

But at least blister packaging and clear plastic bags let you *see* what you're buying, in contrast to the "pig-in-a-poke" method of merchandising, which is also emerging again, for the first time since the thirteenth century. Not infrequently comparatively expensive items, like pocket calculators, come in sealed boxes without so much as a photograph, but with the strong injunction: *"Do not open box!"* Store owners are frequently reluctant to let prospective customers examine the item. Recently when I was trying to buy two expensive electronic parts that had to be compatible, the store manager's permission was needed for me to open one of the boxes to see if the parts fit. When permission was finally granted, the examination had to be accomplished under the manager's coldly disapproving stare.

High-fidelity records, nowadays sometimes selling for as much as $9.95, come in sealed packages, can no longer be listened to at the store and are not returnable if found to contain inherent sur-

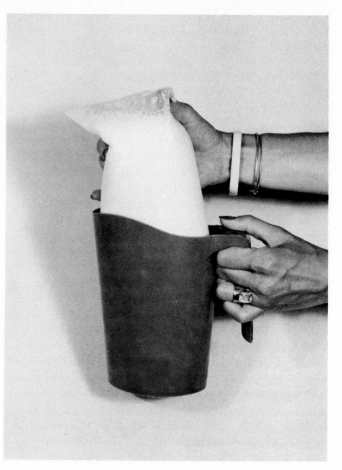

Canadian milk delivery and storage system. 89,   90,

face noise or scratches. Large gift books are similarly sealed (just like pornographic magazines) to prevent examination by the prospective buyer.

We feel that pilferage and damage to the merchandise is now often used as an excuse to push inferior goods. This is particularly true at the supermarket, where fruit is often packed in a plastic-wrapped cardboard tray. The rationale given is that it's more efficient and helps unit-pricing. It is a rationale that holds true only for the store, the distributor and the wholesaler. It also makes it easier in a tray of, say, six oranges to sell four good ones, one that is unripe and green, and one that has rotted. At any rate, who decides that oranges must always be eaten in multiples of six? Or that eggs should come in multiples of twelve?

Not too long ago a dozen eggs came in a molded cardboard box. Now they come in a creaky, pastel-colored, molded plastic-foam carton that interferes with the storage of eggs by cutting off the "breathing" of the egg through its shell. On arriv-

ing home, the eggs must immediately be unpacked (unless one pries the lid open part way). Many refrigerator doors contain "egg holders" that hold nine, ten, eleven or even eighteen eggs, thus always providing either incomplete storage for a dozen or else a continuous reproachful reminder that one really ought to go out and buy more eggs to fill up these cavities. The original cardboard egg package, or its first successor, the molded papier-mâché egg box, was biodegradable, made of recycled materials and not of a petroleum-based plastic.

The latest step in egg packaging, by the way, is a plastic "sausage" containing shell-less, pre-scrambled eggs mixed with such delicacies as grated process cheese and flavoring agents. The plastic tube containing these "instant omelets" is heat-sealed into individual "blisters" of two "eggs" each and costs nearly 50 percent more than the natural variety.

These egg constructs, like most supermarket

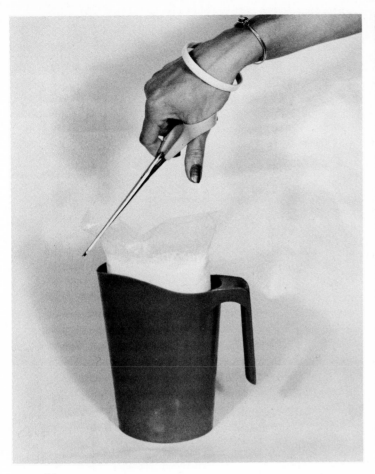

91,

92

luncheon meats, are in the same category of "convenience foods" as the boneless, pale, extruded turkey roll, about as thick as a child's arm and about nine inches long. Until stopped by consumer-protection laws, bacon was packaged in "see-thru" plastic that had been overprinted with red stripes to make the contents look leaner. Sliced ham and other meat products usually come in a transparent blister package, but the contents are nevertheless obscured by a pretty color picture of meat slices. What few consumer-protection laws exist in this area seem rarely enforced.

Frequently packaging is designed to be "aesthetic" rather than functional. Thus one can buy maple syrup in a bottle shaped like "Mother Pallsworthy" (or some other proprietary advertising figure). Unfortunately the good mother doesn't have a decent pouring spout in her head; consequently the syrup dribbles all over "Mother Pallsworthy" and your fingers. Children's vitamins often come in glass bottles that have been expensively molded

into the shape of some cartoon character. In this situation, where the price of the package far exceeds the cost of the contents, guess who pays?

Then there are the gift packages, featuring perhaps a color photograph of a sumptuous Elizabethan banquet, showing minimally king, queen, court jester, a gaggle of lute players and a festive board groaning under the weight of roasted boars, flagons of wine and candlelit cornucopias of fruit. In the package itself is a salt-and-pepper shaker, which can be barely discerned in the color photograph.

That curious confection best called "chip-dip-drip" (and, one assumes, heavily subsidized by the dry cleaners of the United States) is sold in a box emblazoned with color images of improbably distinguished personages living it up in a somewhat languid manner at a cocktail party. Leading felt-tip pens arrive in a box decorated with two nubile girls, presumably "artist's models," who, to judge by their facial expressions, are far gone in ecstasy.

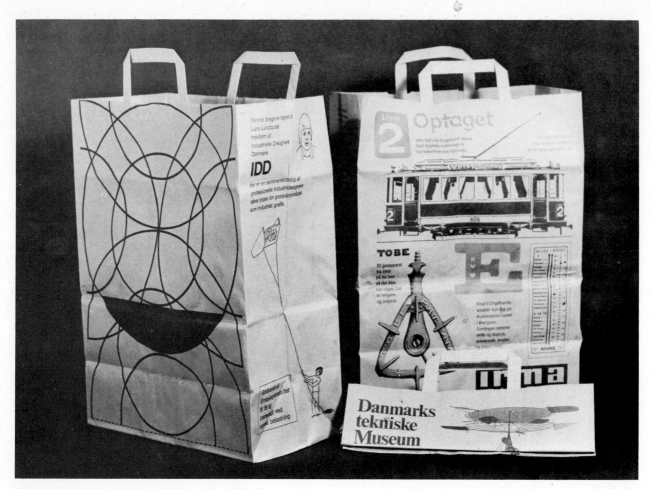

93 Biodegradable shopping bags, redesigned by members of IDD

The box carries no information regarding the pens or their use.

Many packages are restrictive. Welcoming the introduction of "childproof" bottle caps for prescription medicines, one wonders nonetheless how elderly people with arthritic hands or weak fingers are expected to contend with them. Jar lids in general could be redesigned simply, to give the user more ease in opening them. Certainly many a blister package or jar top could defeat Ghengis Khan or Kuth the Impaler.

Tear strips and tear strings are supposed to ease the user's life as well. Band-Aids are packed singly, with a red string that one is supposed to pull clear around the edge to open. Invariably, the string pulls out without opening the package at all. Sometimes the string breaks halfway through, especially on large bulky sacks of charcoal pieces or kitty litter.

Packages are sometimes created to purposefully mislead, regarding their contents. During the late fifties and early sixties a manufacturer of sanitary napkins decided to radically alter the shape of the box. It seems women would send their husbands to the drugstore to buy these items, but men were frequently embarrassed to be seen carrying the all-too-familiar box of sanitary napkins. Hence, the packaging consultants dreamed up the "chocolate-box look." The result was that men were now equally embarrassed to be seen carrying a box of candy.

In the fifties an "executive lunchbox" had to be created. Here the theory was that executives, unwilling (for reasons of status) to carry a workingman's lunch pail, and understandably reluctant to carry a tin box embellished with the image of Hopalong Cassidy or Donald Duck, would be given a well-designed alternative to stuffing their Thermos and sandwiches into the briefcase. The product introduction was handled on so intensive a media-advertising base that the product itself, having been bought by wives for their husbands, was used only a few days. The item was already so well known that people would point, laugh and say, "There goes another of those executive lunchboxes!" defeating the whole purpose of the exercise.

Frequently products are invented and marketed for no reason at all, except to make a fast buck. Besides rocks and air, "People Crackers" for animals are selling briskly because they are "cute" (whatever that means) and because children seem to enjoy animal crackers. (All of this conveniently ignoring the fact that an ever-rising percentage of dog food and cat food is consumed by the poor and elderly, since pet foods provide a mineral-enriched, vitamin-high protein package at comparatively low cost.)

Disposability as the main feature of product-package (or package-product) needs little explanation. Disposable lighters come in vast, backboarded blister packs. These packs are throwaways that do not biodegrade. They contain the disposable lighter, which also is not biodegradable, and in some cases a non-biodegradable plastic attachment into which the disposable lighter can be plugged, converting it to a table lighter.

Some years ago, one could purchase the skeletal outline of a smallish rectangular box, made of aluminum edging. With it went a rectangular transparent plastic bag containing water, tropical fish and enough fish food to last about four weeks. This bag was permanently sealed. The bag would insert in the skeletal support structure: *"instant aquarium!"* After four weeks or so, one could flush the bag, remaining fish food and dying fish down the toilet and substitute a new insert! Hospitals dispose of approximately fifteen pounds of throwaway "linens" and other medical products per person, per day. None of this material biodegrades, although burnable or recycleable paper and cloth substitutes are used successfully in most parts of the world.

☯ Until about three years ago, the Danish chain grocery firm "Irma" provided customers with a strong plastic bag, decorated with their trademark. For ecological reasons the bag was succeeded by a heavy brown paper sack, equipped with paper carrying handles and capable of carrying forty pounds of merchandise. Anyone who has seen elderly people or anyone without a car attempting to waddle along the street with handleless grocery sacks will applaud this carrying provision, considering that paper bags with handles are usually costly and not available at supermarkets. In addition, members of IDD (*Industrielle Designere Danmark*) designed various graphic decorations to be printed on the bags from which children could later create kites, zoo animals and the like (see Figure 93). (This is in the tradition of many American breakfast-cereal boxes, a tradition which has recently unaccountably died.)

Plastic bags can be hazardous; plastic garment bags have asphyxiated children. Paper bags pose no such dangers.

"Second-use" packaging is theoretically a good idea. Again, the "tastemakers" of Madison Avenue have let their Mesozoic imaginations run tame. French mustard crocks, Scottish marmalade jars and German pottery containers for various pickles can be used as baking dishes, serving dishes, pencil containers, etc. The flower-and-butterfly-covered canisters in which coffee, cookies or lard are marketed are all rather nasty. In an earlier book we showed that standard corrugated boxes can be cut and folded into a child's car seat *(if used in conjunction with a safety harness).*[3] In the same volume we showed a second-use alternative for baby-food jars as a workshop storage facility, the use of Styrofoam cups and tin cans for making lamps, and much else. An imaginative company could contribute greatly in this field.

☯ Many bulk items could have second-use designs stenciled directly on the cardboard containers. Laundry detergents and breakfast cereals that come in truly enormous economy boxes could fold and become brightly colored children's furniture.

Labels on packages, jars and bottles ought to be informative. This is especially true of drugs and medicines. In many countries labels on drugs carry only a description of the contents, dosage recommendations and information about possibly harmful side effects.

☯ Since a large number of drugs are used by people over the age of forty-five, and since the accommodation range of the human eye for reading fine print begins to deteriorate at that age, the use of large, easy-to-read type is recommended. (The publishing industry already prints certain perennially reread novels in this manner.)

●

The Arm & Hammer baking-soda people, by contrast, are to be highly commended for the continued use of a traditional package, a trademark familiar for decades and for including on the package useful information regarding the product.

◖ One could raise many more questions: Why don't shoes, for instance, come in a double-divided cloth bag instead of boxes? The boxes are essentially useless (unless a small child wishes to keep caterpillars or frogs in them with leaves and grass). Shoe bags, useful both for storing shoes and packing them when traveling, have to be bought separately as "Shoe-Sox." The questions about packaging are many, the answers few.

The ultimate package is the coffin or casket bought for us after we die. We deliberately did not say, ". . . that we are buried in," since in several states powerful undertakers' lobbies force us to buy a casket even when we are to be cremated and our eventual destination is an urn.

Aside from the ornateness of coffin design that makes them look ostentatious, they are exorbitantly expensive and ecologically unsound, being carefully designed to prevent the natural biological processes from taking effect after burial. Marshall McLuhan in an early book reprints an advertisement that shows a pretty but tear-stained young widow looking out through her window at a rain-soaked landscape, with a somewhat smug simper. The message beneath (illustrating a cross-section through a casket lavishly outfitted with liners, innerliner, plus heavy outside casing): "What a comfort to know that your loved one is dry!"[4]

Even the armed forces, using the ultimate minimal package, a plastic body bag, are careful to divorce the remains from the biosphere.

As we said before, the dividing line between product and package frequently gets blurred. On a British consumer affairs program, a superb example of misleading product design was demonstrated. One of the presenters showed a sleek, low-profile, hi-fi-type AM-FM radio with speaker panels at each end and a multitude of impressive-looking knobs, switches and controls. The set was twelve inches deep, ten inches high and a full three feet across.

After showing the set, the presenter gleefully ripped it apart in front of the camera. The teak exterior was revealed to be stained composition

board. But worse was to come: The total radio behind the thirty-six-inch façade was revealed as a pocketable transistor radio, measuring two by three and a half by five inches. Directly wired onto it were two tiny 39¢ speakers, set at right angles. These in turn fed into two gray cardboard tubes, each about five inches long and presumably originating in the center of two rolls of toilet paper. This device spread the sound (more or less) to the grilles, which were nearly three feet apart. Only three of the controls were connected to anything: on-off, volume and tuning.

This set, made by a British firm, was marketed under a Japanese name, supposedly to communicate Japanese know-how and electronics to the public. Worth about $3.00, the set sold for the equivalent of $95 and, until exposed on television, sold briskly in the British trade. In Figure 94 we show drawings of this type of set both from the front and back. This exploitation of the public was possible because of people's awe of anything that had to do with high fidelity.

Starting out with a good idea, a string of improvements and redesign is frequently added over many years or decades, until the final result is an absolute copy of where the idea originally started out. A good example comes from women's home hair care.

It all began with a pair of curling tongs that were heated over a kitchen range or gas burner in the nineteenth century. Soon these curling irons carried their own gas heater. This, being patently dangerous, led to a plug-in electric version around 1920, and this, still being "inconvenient," led to the introduction of hair curlers in the forties. From early curling devices seemingly made of black chicken wire, the "progress" continued until the late fifties. At that time, sober German citizens around Bremerhaven could still be startled by U.S. Army wives doing their shopping with twelve to sixteen empty, economy-size orange-juice cans built into their hair.

Soon more "scientists" discovered that hair curlers could be directly heated, thus making it no longer necessary to wear them for many hours. However, this frying and cooking process was found to be harmful to hair, thus leading to steam- and mist-producing curling sets. These were soon succeeded by cooking vats for curlers, in which various conditioners could be added to the "soup."

The latest trend toward more natural styles

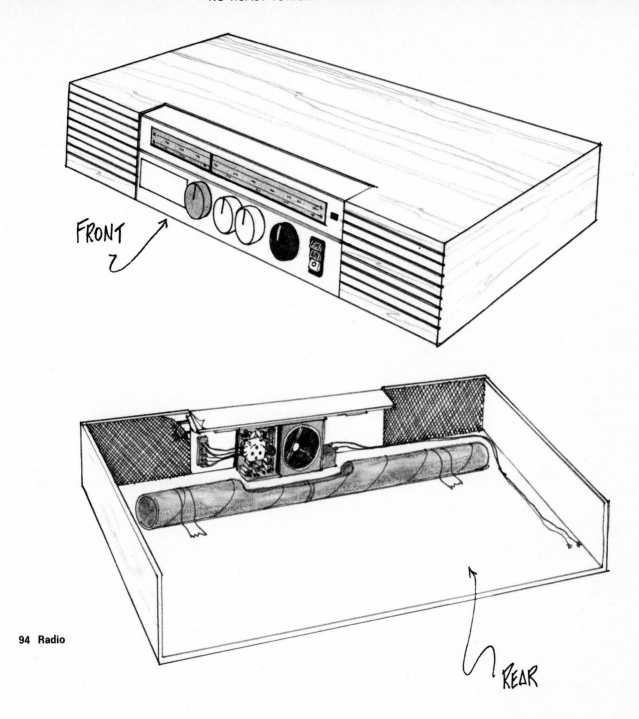

FRONT

94 Radio

REAR

has again outmoded everything. We now have an updated, cosmetically styled version of the electric tongs of fifty-five years ago. The wheel has come full circle.

Many trivial products are still stuck in the in-between category, where additive design has merely made them silly, without as yet coming full circle. Where we mentioned earlier that the lines between product and package frequently become blurred, we now deal with devices that are "service-dispensing" packages masquerading as tools or appliances. The package is the medium.

A frivolous invention some years ago was the electric carving knife. Early models had a fast oscillating blade and were almost vibration- and noise-free. Upon first receiving such a knife as a gift, it would be plugged in and switched on. When seemingly nothing happened, an amazingly large number of people would hold the blade, to "test" it, against their left hand, neatly amputating four fingers. Modern technology immediately worked a hum in, together with some vibration, and the manufacturers adumbrated the box with warnings. More recent exploration has led to electric carving knives with built-in headlights (for candlelight dinners, we presume, and as a further indication that the machine is on).

The introduction of so lunatic a device usually begets a whole string of additional trivia. Shortly after the success of the electric carving knife, some hydrocephalic innovator (read "bubblehead") gave the world the first electric fork. Intended as a gag companion to the knife, it did nothing. The electric line led from the plug to an electrically disconnected strain relief in the wooden handle, sparing the user electrocution.

The electric face sauna (essentially a plug-in steam pot with a plastic shield) of some years ago has been succeeded by the electric face washer. This "skin machine" is a palm-sized, soap-shaped motorized unit that twirls a rubby-dubby brush which, when used in conjunction with soap, is *"essential to deep-pore cleansing comfort."* Mains-operated, battery-driven and rechargeable units are offered for sale. One can assume that both of these devices, as well as many others, are logical (?) extensions of the electric toothbrush and water-pick.

The electric Scotch Tape dispenser quite logically led to the electric stapler, letter opener, pencil sharpener and even the electrically rotated recipe file and telephone index, containing three-by-five file cards. In another direction, the Scotch Tape dispenser swiftly escalated to the electric toilet roll, soon with an electric music box added, which in turn was succeeded by an AM radio dispensing toilet tissue. For the kitchen, an electric unit with four rolls will dispense Saran Wrap, aluminum foil, waxed paper and garbage bags.

Coffee makers also don't work. Percolator types use boiling water, and by boiling the coffee itself they remove the taste. They are also terribly energy-inefficient. The less said about instant coffee, the better. In order to sell less coffee for much more money, coffee now comes packaged as a "coffee bag," similar to tea bags.

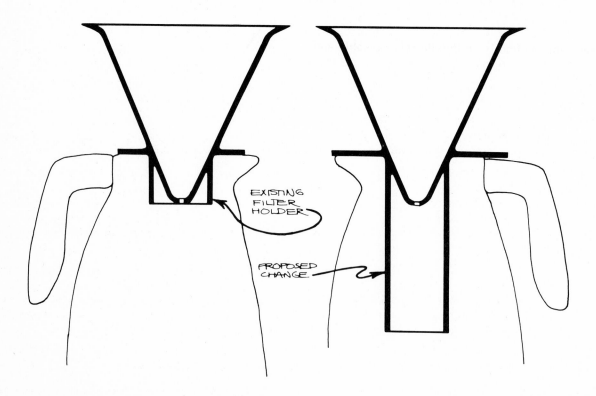

95 Comparison between existing filter holder and our proposed change

The best way to make coffee is by the filtration method. Hot water is poured through finely powdered coffee without ever boiling the coffee itself. Some filter devices work well, others do not. A widely used coffee system is very good except for the separately available, lightweight plastic filter-holders, which are placed on top of a serving pot. The comparatively weighty stream of hot water hitting the coffee in the plastic funnel can dislocate it and spill the contents over a child. Children also frequently knock over or stumble into such a precariously balanced arrangement, spilling boiling water all over themselves. This scalding has led to severe burns and even the death of smaller children. Full statistics with data (and, for those strong enough, photographic illustrations) exist.[5]

⬤ The reason for this malfunction is simple: The lower part of the coffee funnel is kept extremely short in order to be compatible with many different serving pots. If the plastic tube forming this part were to be lengthened by four or five inches, it would all work safely (see Figure 95).

Automatic drip coffee makers usually rely on their own brand of filters, which are expensive and sometimes in short supply. The only exception we were able to find is an indefinitely reusable and washable filter incorporated in the Norelco (Philips) system.

⬤ The best among automatic drip coffee makers is the "Chemex" (see Figure 96), which is designed to be inherently stable. It is a perfect example of what American design can do when it is both innovative and socially responsive. This coffee maker is made of a heavy Pyrex-type glass in a squat, hourglass shape. To reduce the chances of burning one's hand, the grip is made of two wooden pieces, wood being a material that has a good "feel." This grip is held in place by a rawhide thong, adjusted with a cork ball. All natural materials these, pleasant to touch and see. The pouring spout on top of the unit is well designed for its function; all the materials in the Chemex are biodegradable; it will work equally well on a gas or electric stove, an electric hot plate or an open hearth. It is easy to clean and maintain.

The last forty years have also led to an incredible profusion of electric and electronically steered tools for the home craftsman and amateur

96 Chemex coffee maker

builder. They work. But in many cases hand tools still work better. Any good cabinet-maker or craftsman uses hand tools when attempting to achieve extreme accuracy. Automatic processes were originally designed for rapid mass-production techniques. They appeal to generally lazy amateurs, who are more interested in doing a job quickly than well. Therefore an increase in "nonsense" tools developed for such people. A good recent example is the electric drill bit sharpener selling for just under $30. Its job, plus many others besides, can be better performed by a bench grinder with a drill attachment. This also takes less material off, gives greater control, costs the same but ultimately provides greater function because of multiple uses.

⬤ Generally hand tools make it possible to exercise greater skill and because of a slower working rate give greater control over both material and process. They are less liable to deface and destroy materials, as most power sanders do. They give much greater opportunity for improving one's skill; and hence each tool, being an open-ended device, becomes a dynamic learning experience. Finally, manual operation provides greater ultimate satisfaction. We need hardly add that they are inherently simpler, break down less often, are cheaper, are easier to maintain or replace, can be personalized to your own hand, use and waste no power, and pollute less through dust and noise factors.

It would be boring to list any more of the trivia fostered on a public that has been media-manipulated into gullibility. Suffice it to say here that when P. T. Barnum said that there was a sucker born every minute, he could not anticipate our own times when the media boys create millions of suckers every second.

In view of all that we have said, why are people so easily persuaded to buy junk? We can let our friend, the late Alex King, have the last word on that:

Now another of my cherished beliefs is that you can sell the American public anything, if it is properly publicized and sponsored. I firmly believe, for instance, that if somebody put a couple of million dollars into an advertising campaign, you could make the chewing of King-sized goat droppings into a national hobby that would put Wrigley out of business. I can already see the "scientists," in white pajama tops, looking earnestly and scientifically at millions of television viewers. "The Greeks worshipped the goat as a god," they are saying. "The goat is the most ancient symbol of natural virility. Let this powerful symbol carry the burdens of your worries and anxieties. Let it be your Scrape Goat. The purest natural product on the market. It reaches you directly from its original source without adulteration of any kind. The pure goat pearls are packaged scientifically three seconds after they were hatched. Remember, Scrape Goat in the gold-foil wrapper. Scrape Goat for livelier liveliness!"[6]

7

Now That My Hand Is Covered with Soup, Where Do I Get the Cup?

I think no virtue goes with size.

—Ralph Waldo Emerson, "The Titmouse"

In Chapter 2 we advocated sharing and communal ownership. The "public" environment is full of systems, devices, machinery and tools that are developed to be shared. However, they are not communally owned, and in those exceptional cases in which they are, the community is too large in scale to give feelings of personal participation and a sense of connectedness to the user.

In trying to show how things don't work, we have dealt with devices used mostly on a one-to-one basis and small in scale. In talking about the public environment, we need to separate out similar examples, for it is beyond the scope of this volume to criticize the public environment *per se,* or to offer alternatives to it.

That the public environment itself does not work has been fairly well established by architects, planners, social critics and behavioral scientists through hundreds, possibly thousands, of books, articles and case studies over the last fifty years. It need not be repeated here. Therefore let's begin with some of the components that, taken collectively, make up the public environment.

No one has collected statistics on how many people have kicked a vending machine. Nor are reliable figures available as to whether the machine or the person was damaged more. Certainly vending machines form a significant part of the public environment. In many universities and factories, a hallway lined with these coin-operated dispensers has augmented or even replaced the cafe-

teria. The choice of food and drink is a restricted one: Over ten years ago graduate students at Purdue University would show up for class after a refreshing breakfast of a caramel-nut bar covered with milk chocolate, a bag of dry, salted pork rind and a sparkling cup of grape soda. Lunch might consist of a cup of lime yogurt, thinly sliced turkey loaf on white bread, and a carton of chocolate milk. The mind boggles.

The limited and prechosen selection of such foods alone would be intimidating. When coupled with the elimination of the cafeteria, it becomes threatening. For the most part, vending machines don't work. Money gets stuck and is not returned. Cups jam halfway down; the drink pours merrily down the drain. One reason why cups get stuck is that the track gets gummed up with a sugar residue. At the California Institute of the Arts, we watched a student insert a dime in a machine. At first the cup failed to drop and the machine spat a small mound of crushed ice onto the drain grille. Next, a thin trickle of soda saturated the mound of ice. Finally the cup descended into this mess of melted ice and flat soda. The student, unwilling to lose his investment, removed the cup and was last seen walking away, busily licking the remaining residue from the bottom of the cup.

The second possibility provides the cup, but, alas, either nothing to go into it, or else (because the automatic selectors have gone awry) a can of onion soup with extra sugar and cream. Finally, the machine may be out of cups and, as you reach inside, your hand gets drenched with steaming soup.

Automatic food-service machines not only offer a limited choice; this choice is further restricted by the selection made by previous customers: You get whatever is left. Whatever is left is, moreover, highly uneconomical; the machine's prices can be beaten at any grocery store. It must be made clear that this is uneconomical only for the purchaser, otherwise all vending-machine manufacturers and service organizations would have gone bankrupt years ago.

The machines are nevertheless insufficiently serviced. Cleanliness is barely marginal, service personnel desperately attempt to change the merchandise while surrounded by a large group of people clamoring for refunds, food or change. Since the interior of the mechanism must be sufficiently sophisticated to discriminate between various coins and slugs, between cocoa, chicken gumbo and coffee, and a multitude of other possible selections, a large number of controls and servo-mechanisms must be crammed into a relatively small area, making clean maintenance extremely difficult.

The machine depends on electric power. Frequently water has to be piped in, and an electric line heater is attached. When the idiot light blinks on, saying, *"Your coffee is brewing now,"* it really means that water is passing through the line heater, to drip through a miniaturized single-cup version of drip coffee, into the foam cup.

Institutions sometimes place the machine inconveniently or thoughtlessly. In a factory or school, vending machines are often placed within the cafeteria area. The result, of course, is that when the cafeteria is closed, access to the machines is blocked; and when the cafeteria is open, the machines are not needed. The substitution of machines for a cafeteria is also socially unacceptable. It is no longer possible to ask a waiter or waitress for directions, nor is there anyone to whom an emergency can be reported (a fire on the third floor, for instance). Human contact is destroyed. The only friendly face you see during the transaction is your own, provided a mirror has been installed on the front of the machine.

There is a whole second generation of machines whose job it is to "cook" the material emanating from the vendors. At the University of Illinois (when it was at Navy Pier, Chicago), a totally automated cafeteria provided microwave ovens. A hamburger purchased from the vending machines would appear on a paper plate wrapped in cellophane and accompanied, oddly, by a plastic *spoon.* After removing the wrapper, you would place this dish in the oven, select a given temperature and zap away. If the spoon didn't melt into the bun, an indifferent snack resulted. Available from the same facility were chocolate milk shakes frozen into a solid chunk. These, too, were placed in the microwave oven, which melted them to the "right" consistency.

The odd part was that the same temperature setting would be used in both cases! Because of the difference in molecular structure between bun and "burger" (actually a construction woven from soybean filaments), the bun would remain clammy and only partially defrosted, whereas the burger would be sizzling hot.

In spite of these and many other drawbacks to the consumer, vending machines work well for investment capital. It is now possible to buy cigarettes, bathing trunks, condoms, paperback Gothic novels, compasses, pornographic magazines and much else from an ever-spreading horde of inefficient, unsanitary, money-grabbing mechanical robots.

A special place is reserved for stamp-vending machines, placed ever so conveniently in drugstores and motels. Since the recent postal increases, you can buy three 13¢ stamps for two quarters (losing 14¢) or one 13¢ stamp for two dimes (losing 7¢). At drugstore vending machines, you can also buy a combination of a 40¢ special-delivery stamp plus a 13¢ airmail stamp for three quarters (losing 22¢ on each "deal"). The stamp machine then, far beyond Reno and Las Vegas, is the only true one-armed bandit: It doesn't even give you the thrill of a game.

For all these reasons, the individual furiously kicking and rattling a vending machine has become a familiar sight. Vending machines are constantly vandalized all over the country. While we have ourselves frequently experienced rage, anxiety or frustration when faced with these monstrosities, we have only one alternative: *Boycott 'em!*[1]

By contrast we can talk about the telephone as a participatory system and compare telephone systems in the United States with those of Great Britain. In some ways the telephone is one of the most democratic tools. It is a truly "shared" system; until recently no one could own his or her own phone and the telephone instrument in the poorest slum dwelling was identical to one in a wealthy home. Being a leased tool, installation upkeep and maintenance are systems-controlled. New research findings are passed along to users through improved instruments, better service, etc. While the system is quite expensive to each individual subscriber, it is one of the two or three best in the world.

Nearly all we have said above can be put in reverse and will then be true of British telephone services. The system simply creaks with antiquated mechanical switching devices, lines are noisy, misdials and interruptions all too frequent. When something goes wrong, unlike the system in the United States, it takes forever to get service reestablished.

The reasons are not difficult to find. Any large, complex system is fail-prone precisely because it is a large, complex system. Such systems usually contain insufficient feedback. The amount and type of feedback is modulated by your own expectations of the system. In Britain many people have given up on the phone. Others, even in highly placed administrative positions, still consider telephones a "passing fad." Hence reliance on the system is low, expectations are low. All this modulates service virtually out of existence. In the United States, however, we expect immediate attention to phone malfunctions and thus we usually receive it.

Even so, there are still services that lag behind those offered in other countries. In Northern Europe, for instance, it is possible before leaving home to punch in the telephone number of friends you are about to visit. All incoming calls are then diverted automatically to the new number until the command is rescinded. In North America, advanced communications research by telephone companies tends to concern itself more now with data transmission than with the human needs of individual subscribers.

During the last eight or nine years, following several court decisions, the building and selling of "specialty" telephones and telephone-related items have grown tremendously. French turn-of-the-century reproductions and bicentennial telephones abound. This attempt for status through asserting differences in otherwise· anonymous service modules has spread to such an extent that both the Bell Telephone System in the United States and

Bell Northern Research in Canada have now commissioned industrial design stylists to develop their own versions to undercut the competition.

In addition to bicentennial ("I hear America ringing") claptrap, some of the new creations are pseudoantiques. (They are answers to the question: What would a phone look like if designed by Josiah Hepplewhite? Nero? Aristotle or Madame Recamier?) Then there are the futuristic, science-fictional and demi-Gothic (I was a teen-age werewolf on the starship *Enterprise*) versions. There are also electronic models that include digital stopwatches, a push-button-activated call index system, automated grocery lists, calculators, answering devices and much more. Finally there is the minimal-sculpture or kinetic-object approach: A shiny chromium sphere magically opens to reveal mouthpiece and receiver. It may all have started with an article some years ago: "Why should a Telephone NOT look like a banana?"

Our question is, *Why should it?*

There is an interesting corollary to our comparison of U.S. and British telephone services. Postal delivery in Great Britain is fast, dependable and occurs two or three times a day. By contrast, in the United States, postal service is slow, hampered by frequent losses and undependable. There is a direct relationship between the two systems: Americans are constantly amazed by the reliance on letters and postcards in England, whereas the British are bewildered by their trans-Atlantic cousins' willingness to pick up the phone instead. In Britain telephones are frequently vandalized, here it's mailboxes.

Mass transit is one of the great unsolved problems in the public environment. With ever more cars and denser populations clogging city streets and highways, and with pollution factors and the highway death toll constantly rising, alternative solutions seem desperately needed. Vic has written extensively on the subject elsewhere,[2] but then, so have thousands of others. There seems general agreement about the nature of the transport ailment as well as its causes and symptoms. But seemingly there is no cure.

Our alternative thesis is that small parts of the solution are lying around all over the world. They are used and they work. But while an incredible amount of research is going on, many systems that already exist seem unknown to the researchers. The second problem is that presently usable systems, even if known, are not adopted elsewhere. Engineers and designers are apparently unwilling to accept a well-working solution from somewhere else, unless they can make a quantum leap into "improving" and "modernizing."

The Wuppertal Valley in Germany has enjoyed a clean, safe, fast and convenient overhead monorail system for the last sixty-four years. The trains run electrically and are, considering the system's age, not very noisy. Since a series of medium-size cities lie in a belt area throughout the valley, the monorail system replaces private transport between towns and within cities. Engineers and designers elsewhere, though, seem less than eager to duplicate it. Although entering its seventh decade of service, "adaptations" of it usually call for driverless, electronically steered cars and gigantic expenditures. The success of the Wuppertal system, however, is based on modest scale, low operating costs and over six and a half decades of amortization of capital equipment.

To get back to the main point: Since partial solutions are being used in many parts of the world as good alternatives, what is needed is synthesis of all these various subsystems into a well-articulated overall mass-transit system. Some rules for mass-transit design can be summed up simply:

1 *Try to use what already exists in an imaginative way.*

The Canadian government, in spite of rising gasoline prices, is seriously considering phasing out both of its rail systems. The United States has abandoned hundreds of thousands of miles of rail network, including spurs reaching into many tiny towns, especially in the East, South and Midwest.

2 *Look around at other possible systems, both to prevent failure as well as to adapt other technologies.*

Despite the monumental failure of Charles De Gaulle Airport near Paris, the Canadian government has constructed a new International Airport called Mirabel forty miles outside of Montreal. International flights from the existing Montreal airport, Dorval, have been eliminated. Mirabel is about an hour's bus ride from the center of Montreal. It is two and a half hours by bus from Ottawa, and only six flights a day connect the two cities. The airport has been established at staggering costs, directly in the worst-hit part of Montreal's snow belt. It is situated on land that is low-lying enough to be badly affected by frequent fogs. Few flights use the facility, a great deal of traffic and passengers preferring to use Toronto, or Kennedy in New York.

With these two lessons in mind, one wonders at the mammoth Dallas–Fort Worth airport, built for $700 million. This facility is located halfway between the two cities, and a $14 cab ride from either place. Minor services (going from one airline to another, a local phone call or a cup of coffee) cost 25¢ each. Change machines return only 95¢ for each dollar bill presented. The biggest problem is the much publicized "Airtrans," a thirteen-mile computer-controlled train system that shuttles passengers around the airport. Trains frequently skip stations, or fail to open doors after stopping, and passengers routinely miss flights while trapped in the cars. An overcomputerzied baggage-handling system has destroyed or chewed up one out of every eight bags handled. Several airlines have moved part of their operations back to Love Field, others are talking of moving out permanently.[3]

3 *Try to use working technology as it operates now, and not anticipate high-technology futuristic development.*

(The landing mast for zeppelins atop the Empire State Building.)

4 *Keep the system on a modest scale.*

The STOL airplane shuttle connecting the downtown business center of Montreal to a central location in Ottawa worked extremely well, with more than a dozen commuter flights daily, during its trial run of nearly three years. When time came for the program to be reviewed, the government scrapped it, since it offered no possibilities for expansion, *even though it was demonstrated that it could continue to serve well indefinitely at its present size.*

To repeat: A well-articulated mass-transit system would consist of short-range, medium-range and long-range transportation devices, the first of which would also have to pay attention to inner-city needs. Designing such a system merely entails plugging together what already exists in a logical manner. Here are some of the systems now in use:

● In Amsterdam the *Witkars* ("White Cars") are beginning to revolutionize inner-city transportation. *Witkars* look like squat white telephone booths on tiny wheels. They are two-seater, drive-it-yourself electric vehicles that can reach a top speed of twenty miles per hour. They can cruise between carefully spaced stations in the inner city that are less than two and a half miles apart. The car can be recharged in five minutes. There are several dozen in Amsterdam now, and eventually 1500 of the nonpolluting and ultracompact cars are expected in the city.

The Witkar Cooperative Society already includes more than 1500 members who pay $15 each for a key that unlocks the car from the recharging station and, at the same time, signals a computer that begins charging the account of that particular driver at the rate of 3½¢ per minute. This is less than half the taxi fare for a standard trip. Five cars occupy the same amount of space as one American automobile. While the cars created a sensation in Amsterdam on their introduction over two years ago, by now they attract hardly a glance, except from tourists.[4]

● In several smaller cities in Denmark, stacks of free bicycles painted a distinctive white are available for inner-city transport. Anyone can take a bike and ride it anywhere within the inner city. Because of natural traffic patterns asserting themselves, the bicycles tend to collect around department stores, bus and train stops. They are thus always available where needed most.

● The city of Bologna in Italy, with more than half a million inhabitants, began charging no fare during rush hours on all public transport, starting April 1, 1973. Rides are free up to 9:00 A.M. and again from 4:30 to 8:00 P.M. During more than three years, buses showed 8 percent greater use during the morning and 30 percent greater use in the afternoon. Traffic in the inner city flows more smoothly and safely, and large downtown areas are completely free of private automobiles.[5]

● In enlarging the international airport at Copenhagen, the SAS airlines system considered the installation of such exotic and high-energy-reliant devices as moving walkways or "people movers." Instead, it was decided to use nonpowered scooters, which are left at the various departure gates so passengers can scoot themselves and their carry-on luggage back and forth. Since the system does not work well with elderly or obese people,

97 "Spillproof" scooter, designed by Tim Lloyd

one of our students, Tim Lloyd, has designed an improved version that is nearly spillproof (see Figure 97).

● In 1975 Lucas Industries introduced a subcompact minitaxi that is electrically powered and intended for inner-city use in London. (See Figures 98 through 101.) The vehicle had to conform in every way to the rigorously enforced criteria of Scotland Yard's Public Carriage Office. The first prototype model is the result of two years' work. The superstructure of the taxi is glass-reinforced plastic, with a steel roll-over bar as a divider partition and door pillar. There are GRP corners that can easily be replaced in the event of minor accidents. The taxi is thirty-nine inches shorter than conventional London taxicabs, so that five could stand in a rank where only four can presently park. Seating is as comfortable as in standard London taxis, retaining the same unusually high head clearance.[6]

On June 18, 1976, the Museum of Modern Art in New York City opened an exhibition under the

The new London taxicab. 98,
Photo Peter Baistow. Courtesy *Design* Magazine, England

99,
Photo Peter Baistow. Courtesy *Design* Magazine, England

100,
Photo Peter Baistow. Courtesy *Design* Magazine, England

101
Photo Peter Baistow. Courtesy *Design* Magazine, England

102 **Maintenance and repair barge-"train" for British Waterways canal system. It consists of a series of barges that will house a working party as well as carrying special-purpose boats. Designed by David Raffo, a postgraduate student at Manchester Polytechnic.**

title "The Taxi Project: Realistic Solutions for Today." Five taxi prototypes, each unusually "clean" from the viewpoint of air pollution, were exhibited. The taxis also are roomier and have comfortable accommodations for people in wheelchairs and women with baby carriages or strollers. Steam Power Systems of San Diego and the American Machine & Foundry (Advanced System Laboratory) of Goleta exhibited steam-powered functioning prototypes. Volvo showed two models (on which we had done some consulting) that are again receptive to people in wheelchairs and stress security for the passengers. The models shown by Volkswagen featured a hybrid power plant, which is a combination of an internal combustion engine and an electric motor (it produces no exhaust at all when operated on the electric plant alone). The Volkswagen models also accommodate several people in wheelchairs, are heavily noise insulated, and carry an intercom unit for medical and traffic emergencies. Alfa Romeo presented their version, which is based on a normal internal combustion engine but again makes special provisions for people in wheelchairs and with baby carriages, in sketch form only. After the exhibition the prototypes were tested by the New York City Taxi and Limousine Commission and will be used to set standards of emission control, economy, adequate and comfortable accommodations for passengers and luggage, safety, maneuverability and driver comfort, which all taxicabs must adhere to in order to be approved for use in New York City after 1977.[7]

In West Germany, taxis have not been permitted to cruise around looking for fares since the oil crisis of October 1973, and because of a growing awareness of air-pollution factors. Instead, lamppost-mounted call boxes exist at all inner-city and suburban rail stations. Pushing a button, the location of the caller is transmitted to the nearest taxi rank. A taxi can be guaranteed to arrive within not more than two and a half minutes.

In Ottawa, Canada, as well as in many cities in the United States, dial-a-bus systems of various sorts are in operation. Because of Ottawa's dreadful winter climate, the local system works this way: You phone for a minibus, which will pick you up at your home within five minutes during peak hours. The bus will make its rounds and deposit you at the nearest bus shelter, where a "regular" bus can be taken to a scheduled destination. From there, a second minibus will take you to the house, business or shop you choose to go to.

The Wuppertal monorail system has already been mentioned. Other inner-city rail links deserve mention. In England hundreds of commuter trains take people to work and back. The train cars, while many decades old and in a generally run-down condition, nonetheless feature the "breakfront" concept—that is, every eight seats have their own door so that a carriage can be emptied or filled with commuters in seconds. The rail system terminates at major underground stations. In Copenhagen more modern "S-Trains" connect the inner city with all suburbs, major rail links and bus stops. While transportation in Copenhagen is not free (as in Bologna), it operates essentially on an honor system.

For city-to-city transportation there exist high-speed as well as extreme high-speed "bullet" trains in Japan, West Germany and France. STOL Air Traffic in Canada has already been mentioned, and there is some limited VTOL traffic in Eastern Europe.

For water, or water-and-ground, transportation, there are high-speed hydrofoils in continuous use in Finland, the Scandinavian countries, Russia and Eastern Europe. Hovercraft automobile ferries cross the channel between France and England several times daily.

⚫ For long-range transportation two vastly differing directions can be followed: One is the fast aircraft route, eventually terminating in supersonic planes. For less energy-consuming and less polluting traffic at lower speeds, the reintroduction of zeppelins, as well as the use of computer-guided servomechanisms that help one or two people to steer vast sailing ships across the oceans has been suggested. At the present time, research laboratories in Great Britain, West Germany, Sweden and the United States are making such systems possible.

⚫ Finally, there are the vast existing transportation networks, falling into disuse or facing complete phase-out. With the introduction of high-speed transport, the corollary: Low-speed and inexpensive transport becomes important again. In the northeastern United States a once-splendid canal system reaching parts of the Midwest is either not used at all or used by pleasure craft only. England is crisscrossed by a superb canal network that grew as a means of draining off water from eighteenth-century coal mines. This system, designed for transport by "narrow-boats" (nine by sixty-four feet), has been allowed to silt up or decay in a number of places (see Figure 102).

⚫ The United States is also crisscrossed with disused rail lines. The rights-of-way are still held by railroads or holding companies. As a theoretical traffic network it makes sense, regardless of whether rails are still in place or not. In most cases the original overpasses, underpasses and viaducts still exist.

⚫ It has been suggested that where rails are still in place, trolley systems might be installed with minimal effort.

⚫ Where rails are gone, one-way traffic can be effected using buses or minitaxis, or else bicycle paths, and cross-country skiing trails might be opened up as well.

⚫ If properly paved, two-way traffic, using buses or trolley buses, could be established. All this could be done with little effort and, more important, without dislocation of people or homes.

To return to the public environment. Some objections are that it's ugly, noisy, that it smells bad and that there is a lot of junk in the water and in the air. But such statements are nonquantifiable, and because of their very vagueness they easily lend themselves to inflamed rhetoric and sociological jargon. What is most needed are facts.

⚫ We advocate the use of simple diagnostic devices with which ordinary, untrained people can determine decibel noise levels, particulate air-pollution factors and other vital measurements. These would also include lighting factors in lumens, which must include underlighting and overlighting that can result in eye strain, headaches, injury and fatigue.

Such devices now exist, but are expensive and frequently overdesigned and hard to learn, hence they are restricted to a technical and scientific elite. They could, however, be redesigned easily to come as do-it-yourself kits that could be built by individuals or by groups of concerned people. This would give workers themselves ways of determining some of the causes of their high anxiety levels. Also, a group of people living near a factory could provide its own air- and water-monitoring stations.

⚫ In fact it can be even taken a step further. The inhabitants of Dyrup (a small village near Odense, Denmark) have compelled a local factory to build air-monitoring stations in a ring around the factory. Townspeople monitor pollution according to government environmental standards and can deal directly with the factory or local government to restrain abuse. We feel that if the labor unions concerned with work safety, and ordinary people concerned with airport noises, pollution factors, etc., were to apply pressure, an organization like Heathkit could easily supply their demands for diagnostic kits.

Social concern has become socially acceptable. It can be shown that because of awkward placement and the general amount of effort needed to effect change in our public environment, the man-made environment discriminates against people who can loosely be termed "handicapped." Elderly people, pregnant women, small children, people in wheelchairs and on crutches, women wheeling strollers, prams and shopping carts are all faced with obstacles. These include public lavatories, phone booths, lockers, turnstiles, ticket booths, change-making machines, stairs, escalators, revolving doors, elevators and much else. Public awareness of the legitimate needs of these groups is high enough now, so that we hope for a continuing reexamination of these obstacles placed by the few in the paths of the many.

8

⊙ You Too Can Own
a "Handyman's Special"!

Garbage is the only raw material that we're too stupid to use.

—Arthur C. Clarke, *Profiles of the Future*

There can be bitter and sustained arguments about the "Standard of Living" vis-à-vis the "Quality of Life." There is no doubt that in North America we have achieved the highest level of material possessions. Many of our problems, however, seem to stem precisely from that. Our homes are glutted with products we infrequently or rarely use. Compared to the rest of the world, we seem to be possessed by an incredible number of things. Brenda and Robert Vale put it this way:

> We are persuaded to expect a higher material standard of living when, for the majority, the standard that we already have in the West is perfectly adequate. A marginal increase in this standard can only be made with the use of yet greater quantities of the existing resources of the Earth. What are essentials for the American Way of Life (full central heating, air conditioning, a car per person) are considered, although less so now, as luxuries for Europeans and what are considered necessary for a satisfactory European life (enough to eat, a home and fuel to heat it, access to transport) would be luxuries for the "Third World."[1]

Living among this overabundance of manufactured objects, one is tempted to think that there may already be more than enough of everything. One of our friends has a car radio, an AM-FM plug-in digital clock radio with alarm to awaken him in his bedroom, a pocket-size transistor radio, an AM-FM stereo tuner as part of his hi-fi rig, a kitchen radio with a built-in clock and intercom, a radio/record player in the kids' room, an old table radio in the garage workshop, a radio with cassette player and CB transceiver in his pickup truck and a twelve-band shortwave plus AM-FM portable for picnics. Incidentally, he never listens to any of them; instead he watches ABC's "Wide World of Sports" on one of his four TV sets. But the statistics remain: nine radios and four TVs for a family of three. To this *could* be added his "Weathercube," which stands on his office desk emitting constant weather reports at the push of a button.

Similar overpossession patterns could as easily be established for other middle-class families. These patterns generally begin with a person first buying the basic product. Next comes the portable version, and after that the battery-rechargeable or cordless one. Finally, a newer version of the original product, with the more antiquated one destined for the kids' room or, if it is something like a kitchen blender, a box in the garage. And then there were four . . .

As the boxes of unneeded objects fill to the bursting point, more boxes are added. This ends, as can be seen in every tract and development area throughout this country, in cars being permanently parked *outside* the garage, since there is no longer room for an automobile inside. The harbingers of spring are, first, National Groundhog Day; somewhat later, the first appearance of a robin in the backyard; and finally, the garage sale.

101

In late spring, garage sales multiply geometrically, for this is the time when people pack up and move. Just before the moving truck pulls away there is a desperate last-ditch attempt to get rid of toreador pants, slide rules (superseded by pocket calculators), washers, dryers, refrigerators, stoves, baby furniture, tricycles, old machine tools, cross-country skis, pottery, knickknacks (including velveteen cushions with the hand-stenciled message *"For You I Pine and Balsam"*), garden gnomes, reflecting balls, flamingo-enriched screen doors, prom dresses and bowling balls.

In a society stifled by its own possessions, there are a number of avenues for getting rid of things. Besides the garage sale, there are an increasing number of flea markets, swap meets and home auctions. Things for sale can be advertised on supermarket bulletin boards, company or university bulletin boards and in school or factory newsletters. Then there are the "For Sale" ads in daily newspapers and, more recently, a whole series of neighborhood-circulated, free catalogs of things for sale, under such names as *Pennypincher, Pennysaver, Junquetique News.* Finally there are similar newspapers, consisting entirely of for-sale ads, that are sold in neighborhood stores under such titles as *Swap Sheet* and *The Garage Sale.*

Special interests are also served, by such magazines as *The Antique Collector, The Tropical Fish Fancier, Road and Track, The Mart* (from the National Association of Watch and Clock Collectors), the weekly *Antique-Traders Journal, Model Railroading News, Sports Car and Driver, Hang-Gliding Parade, Fine Woodworking,* plus many others devoted to pets; stamp collecting; used beer cans; Mickey Mouse and Coca-Cola memorabilia; guns; Nazi uniforms, insignias and weapons; scuba diving; photography; etc.

Another excellent source for "buying used" is a university town in late May and early June. Graduate students, who may form the most stable part of our population (living in one town for six to ten years), break up house and move. Major appliances like refrigerators, stoves, washers and dryers sell very cheaply and abundantly, the more so since the possible buying population (other students) is also moving away for the summer. Conversely, the same location is a good place for selling things in early September.

Frequently the description of a used appliance, or automobile, carries the accompanying phrase "Handyman's Special" or "Mechanic's Delight."

While this can mean that the car no longer has an engine, or that the washing machine has a burned-out motor, it more frequently means merely that the casing on the vacuum-cleaner head has a crack in it, or that one wire in "Mr. Coffee" has worked itself loose. People have become more and more unwilling and unable to repair even the simplest malfunction or defect. Throwaway tissues have paved the way for throwing away appliances when they stop working, have a cracked base or even just a chip in the handle.

In this chapter, we hope to give you a basis from which you can buy used things with confidence. We will give you simple directions for inspecting what you propose to buy and, once purchased, to diagnose faults, repair and maintain the object.

The first step before buying something used is to determine whether what you are proposing to buy should be bought used at all. Secondhand merchandise can offer the greatest savings, but first you must establish your criteria. A list of criteria should include the following:

1 *Certain high-technology products are impractical to buy used.*

Electronic components have parts that are permanently encapsulated or sealed, and thus cannot be inspected or repaired. Advanced high-fidelity equipment tends to be beyond the expertise of seller, buyer or even the neighborhood repairman. We feel that within the hi-fi field, only speakers can safely be bought at garage sales. You should never buy an inoperable microwave oven, since repair might result in health hazards caused by emitted rays. With most recent high-technology equipment, dealers usually give generous trade-ins; hence if such an article shows up, it often means either that it doesn't work properly or that it is stolen merchandise.

2 *Other equipment may be outdated, parts may no longer be available or only available at great expense.*

Tube-type radios, hi-fi equipment and TV sets fall into this category. Many manufacturers are required to maintain parts services for only five years after the product was originally sold. Thus, a safe rule would be not to buy anything needing parts that is more than five years old, unless you are unusually handy and can fabricate parts yourself, or know of a source where parts are squirreled away.

A certain personal knowledge of the product may give you enough confidence to knowingly violate this rule. The excellently designed 1936 Leica III-C camera will still be serviced by the factory, and parts are still made. This is true also of early model Rolleiflex cameras, as well as old Bolex and 16mm Bell and Howell movie cameras. It is not true of *all* cameras; especially, "regular" 8mm movie cameras are impossible to maintain because of the introduction of "super-8."

Then there is the matter of personal choice. A $35 slide rule (an obviously outdated object) may now be picked up for a few pennies at a garage sale. If you choose to work with a slide rule rather than a pocket calculator, go ahead and buy the slide rule.

3 *The device may not work safely or efficiently.*

A number of electric appliances that were developed years ago still work properly, but are hazardous compared to newer versions that incorporate recent safety innovations or safer assembly procedures. Older top-loading washers, for example, do not have safety cut-off switches that interrupt the spin cycle when the lid is raised. Much electrical gear may come with frayed power cords, which can be replaced easily and then operate safely again, but electric hand tools made prior to 1970 are usually not insulated and may create a shock hazard. The buyer must have a general awareness of safety considerations.

4 *Some objects, because of their very nature, are rarely offered secondhand unless they have been overused or abused.*

High-performance cars, and those pretending to be, are often pushed beyond their capacities. In addition they have often been "souped up" or tinkered with to the point of absurdity. You should beware of buying an innocently advertised automobile which, on inspection, turns out to have racing stripes attached from front to back on doors, roof, hood and trunk; a raised rear end with Wide Oval slicks; an engine compartment full of glittering chrome parts; jewel-bedecked mud flaps; an angora-fur-lined interior; multiple tail pipes and glass-lined mufflers that sound like a mix between a Lear jet and a road grader.

One would also beware of purchasing an electric guitar with high-wattage amplifier from an acid-rock musician, or a "slightly" used lawn mower from a professional landscape gardener, or a family sedan that has the word "TAXI" neatly painted over—or a hole in the roof where a cop's rotating red light used to be mounted.

5 *The nostalgia craze has brought many articles into prominence and driven up their prices.*

Until recently, at swap meets you could always find a shoe box full of old electric hair tongs selling for as little as 50¢ each. But with the hair-care-cycle having come full circle (see Chapter 6), they now frequently sell for $10 or more. Our point is that you can get a brand-new one, with unfrayed cable, for $6.98 at your discount drugstore. Especially in the East, Navajo jewelry of inferior quality sells at swap meets for prices far exceeding those fetched by first-rate pieces from Indian traders (who themselves already charge absurdly high prices).

By contrast, we can recommend certain things almost without reservation. Large, bulky, heavy-to-move furniture and built-in banks of kitchen cabinets or seating units are often ridiculously cheap because you are expected to dismantle them and almost always required to transport them yourself. Some of these things may even be given away free if you are willing to take them away. This category includes pianos (there's a comparatively small market for them, they are hard to handle, and first-time users mistakenly anticipate frequent tuning expenses) and above-ground swimming pools (when the plastic liner is beginning to reach the limit of its useful life).

The secret of avoiding frequent tuning of the piano is simply this: Moving it tends to twist the frame, and changes in temperature and humidity further kick the piano out of tune, so when the piano is first placed in the new location, it *will* be out of tune. People then make the mistake of bringing in a piano tuner immediately, before the instrument has had a chance to become acclimated to the new conditions. We suggest that you be patient for a month to six weeks, *then* call the tuner.

As to replacing the liner on a swimming pool

you have just carted away, we strongly advise that you go back to the original manufacturer and, working with the specific installation manual, put in the new pool and liner according to instructions rather than just "slopping it in."

Some things inherently increase in value on the used market. Besides antiques and nostalgia items, there are such collectibles and "investments" as firearms, stamps, coins and some early toys. With present-day deterioration in workmanship and materials, we can anticipate price rises on the used market for such basic kitchen tools as pots, pans, meat grinders, chopping blocks, tea kettles, etc. One can further speculate that items seemingly outmoded (like the slide rules mentioned above) will appreciate in value over the next decade or so.

Having established that you can confidently buy some things from a garage sale or a "for sale" ad, we can now develop a list of guidelines for mechanical and electrical products:

1 *Ask to have the item demonstrated.*

If it works the first time, chances are it will continue to do so. Listen for unusual noises or excessive vibration. If accessible, feel the motor housing for high temperatures.

2 *If it cannot be demonstrated to you, it is too risky to buy.*

You may come to the house and find the gleaming washing machine in the utility room. To your request for a demonstration the response may be: "Well, I'm not really the lady of the house and I don't want to tamper with it," or you may be told that there are no dirty clothes or detergents available. Walk out.

3 *If the appliance is too far from electric or plumbing connections, be suspicious.*

Remember that this is the way most appliances are presented to the buyer. The disconnected dishwasher will usually stand in the garage with the explanation, "We've bought a new one." But suspicion alone should not deter you from considering the purchase, as some of the best bargains are to be found this way. Actually, an appliance standing in the garage can be much more closely inspected (see guideline 4 below) than when it is still installed under the sink. After having checked it visually and man-

ually, you can then ask to plug it in, briefly, so as not to burn it out, as a last check on the motor.

4 *Inspect the exterior and interior of the appliance closely.*

First look at exterior panels, gauges, indicators and controls. Missing knobs are unimportant; they can be replaced easily and cheaply. Superficially you should look for scratches, dents, cracks or other indications of mishandling or abuse. Also look for seepage stains or rust near the bottom. Look for a frayed power cord or damaged hoses.

Now look at the rear inspection panel. First examine the screw heads affixing the panel for signs of excessive wear. This would indicate that the panel has been removed many times for repair. Then remove the panel. Look inside and inspect the mechanism closely. If any part looks newer than the rest, it probably has been recently replaced. On a ten-year-old machine, a new pump would be a good indication of maintenance by the owner. On a relatively new machine, however, replacement of a part may be a clue to serious malfunctioning.

Look for a deposit of metal dust at the bottom of the machine, a sure sign of excessive wear. Jiggle the drive belts to see if they fit snugly. In the case of a washing machine or dishwasher, you should look for interior signs of water staining or rust. Sometimes you may be told that the machine or appliance "just doesn't work anymore." If you then see that the interior of the machine is covered by a uniform film of dust, you can safely assume that the owner has not maintained it properly since it was bought.

A certain degree of buying savvy now enters the picture. Ask the seller questions about the appliance to see if he seems knowledgeable about it. The general condition of the home may give you an insight as to whether the people you are dealing with are careful about their possessions. If you are left in doubt, forget it.

5 *In spite of the pejorative tone of some of the above, most people are trustworthy and open.*

If what they are trying to sell is really worthwhile, they are willing to demonstrate it, and in fact, will bend over backward to discuss the moods, eccentricities and habits of the machine. Frequently they will even brag about the way it works.

Let's assume you have made a purchase. You are now faced with problems of maintenance and, after a while, problems of repair. (Repair problems and maintenance procedures, of course, have to be dealt with even if you buy everything new.)

The repairman's role is deteriorating. Since he deals with subassemblies, which he unbolts and ships back to the factory for replacement, his role is often reduced to a diagnostic one. Misdiagnosis can and does occur. Sometimes you may be told that an appliance is beyond hope, yet you yourself may be able to locate the trouble spot and effect a simple and inexpensive repair. Nearly everyone has heard stories where someone was able to repair something that had been dismissed by the "service representative."

To repair something yourself, you need to gain knowledge of the product and how it operates. There are generally two ways of acquiring this knowledge. One is to insist on the specific repair manual and owner/operations manual when you first buy the device new or used. When buying something new from a shop, you must insist on this and refuse to purchase if the product is unaccompanied by such instructional material, or unless the manual must be requested by mail from a manufacturer. On electronic devices, high fidelity, etc., you should also ask the seller for the schematic diagrams. Often a circuit schematic will be glued to the bottom of the case.

If no manual exists, or if the instruction booklet for the product you have bought used has been lost, there exists a huge library of maintenance and service manuals for all conceivable devices, although most of these are somewhat more general than the specific pamphlet accompanying an appliance. The manuals can be bought or consulted in the library, and should be studied and read until you feel comfortable with the product described.

The second method is to take evening classes. These are offered in such subjects as appliance repair, automobile maintenance and repair, television repair, and home-building (including plumbing and electrical work). In a few cities, courses on special and generalized maintenance and repair are being offered. But don't expect to be able to take a special course on how to fix your "Astro-flite" toaster—courses are never that specific. Instead, try to obtain a specific manual. What we are advocating here is that you take courses dealing with maintenance. Other courses exist (including correspondence schools) that are designed to prepare you for a career. Such courses are usually expensive and far too technical for your limited needs.

Building an appliance from a kit (see Chapters 3 and 4) is another way of building up your knowledge and ability to maintain, diagnose and repair as well as having on permanent file all the appropriate manuals.

If the device has stopped operating, there are sequential steps you can go through to get it going again. Having gained knowledge from reading technical manuals, you may be able to simply repair it. If the device is quite old and not very valuable, you may just want to "dive in" and try to repair it, gaining knowledge and understanding as you go on. Next, you might wish to seek the help of a friend, a talented neighborhood youngster or a local "fix-it" guy. All else having failed, and as a last resort, you might take it to a shop. However, all of the above have usually enabled you to know at least *why* the machine has stopped. Minimally you have acquired a terminology of troubleshooting, and are capable of reeling off such phrases as "The centrifugal clutch is slipping on the washing machine," thus indicating knowledge. This alone may save you from having repairs performed that are not needed, or otherwise being swindled.

Once you have become electrically proficient, you will be interested in simple devices that will assist you in diagnosing and repairing your appliances. Figures 103, 104 and 105 show two such devices. Both are very inexpensive. The Outlet Tester is ultra-simple to build and will tell you the condition of any wall outlet in your home. (Before you take any appliance apart you should always check its wall outlet first!) The Appliance Tester is equally simple to construct and exceptionally informative for analyzing major appliance faults.

In our view, Sears, Roebuck and Company offers the best backup service available for appliances and other products. Sears seems willing to exchange and replace without question products that are shown to be faulty. Their nationwide chain of stores makes it possible to always be close to one of their locations even if you have moved a considerable distance with your appliances. They

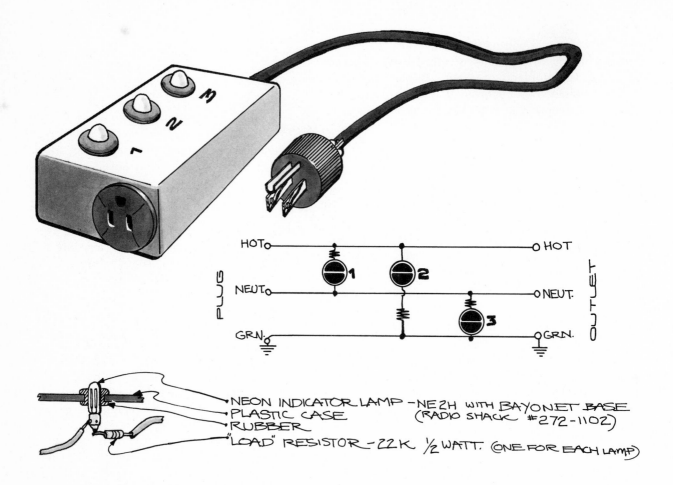

HOT ○ ─────── HOT
NEUT. ○ ─────── NEUT.
GRN. ○ ─────── GRN.
PLUG · OUTLET
1 · 2 · 3

NEON INDICATOR LAMP — NE 2H WITH BAYONET BASE (RADIO SHACK #272-1102)
PLASTIC CASE
RUBBER
"LOAD" RESISTOR — 22K ½ WATT. (ONE FOR EACH LAMP)

103 Outlet tester. This circuit can be built into any suitable box. Lamps are mounted by being pushed up through a rubber grommet (see detail above). Now plug tester into an outlet:

lamps 1 & 2 light up:	outlet is normal
none light up:	open "hot" lead
2 only lights up:	open "neutral" lead
1 only lights up:	open "ground" lead
2 & 3 light up:	"neutral" lead is "hot"
1 & 3 light up:	"neutral" and "hot" leads reversed

NOTE: Any condition other than lamps 1 & 2 lighting up is dangerous and should be corrected immediately.

have repair centers that stock most common replacement parts and specialized tools. They offer repair services, as well as a great deal of help across the counter to perform your own work.

In addition, even the smallest and most remote Sears catalog center has a microfiche reader. They have files of microfiche cards, showing blowup drawings of every appliance and product they make and sell. These can be scanned and the needed part's number determined. After ordering,

the components required for the repair can be in your hands usually within a week. Moreover, every Sears tool and appliance comes with a complete owner's manual and blow-apart diagram, showing every component and all part numbers. Hence you can order parts directly from Sears, based on your manual.

Since this far-going backup is official Sears policy, their own brand of tools and appliances tends to be sturdily built and put together in a logi-

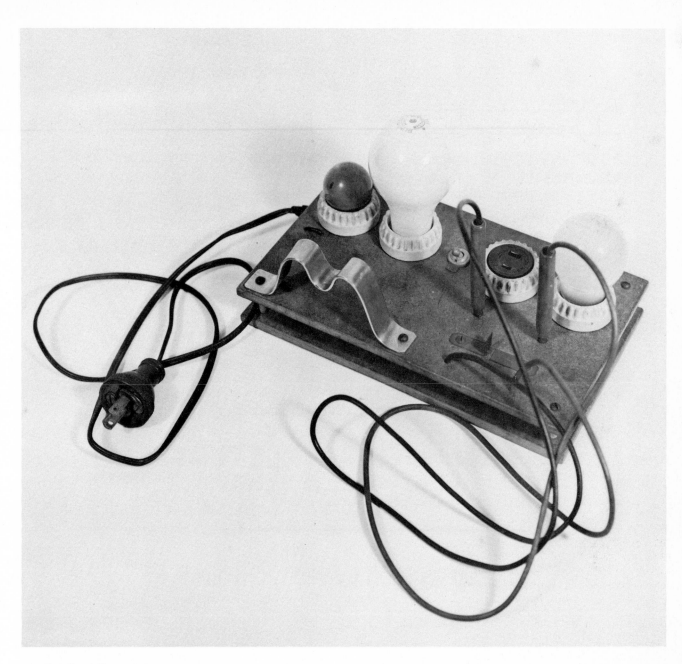

104 Simple appliance tester

cal sequence so that repairs can easily be made by their own repair center or by the customer. For the same reasons, Sears warranties and guarantees, both implied and explicit, are generally broader than those offered by many other manufacturers.

Generally warranties differ greatly in what they define as "normal use." A leading hi-fi manufacturer expects "normal use" of the turntable to consist of playing less than two hundred records a year. This number is far below average usage.

One of the best-known makers of Japanese precision cameras defines "normal use" as pushing thirty or forty rolls of film through the camera each year, a number easily equaled in one day's shooting by the many professional photographers who use this camera.

Again, it is beyond the scope of this book to discuss the fine legal technicalities of warranties and guarantees. There are, however, a number of totally phony bits of "consumer protection" floating

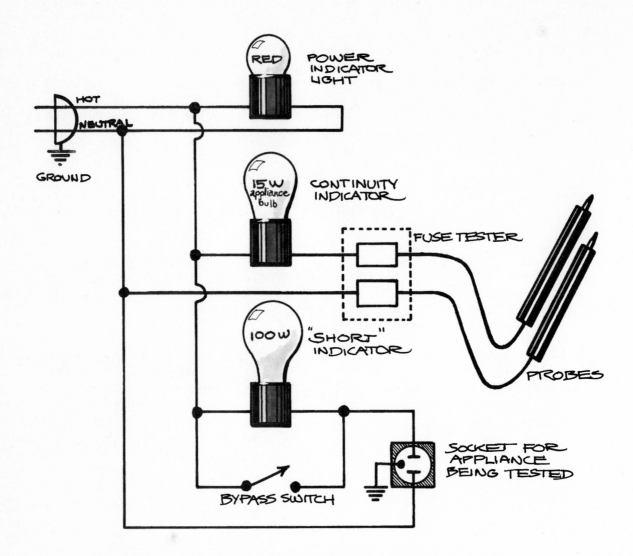

APPLIANCE TESTER CIRUIT

105
1. Test outlet for correct polarity and wiring (see illustration 103)
2. Plug in tester; red light should go on, indicating the presence of power
3. Test for "open" wires and fuses with the continuity indicator
4. Test appliance by plugging it into the test socket (make sure by-pass switch is in "off" position). The 100-watt bulb should light dimly. If it does not light, the appliance is "open"; look for a broken connection or loose wire. If the lamp lights brightly, check for a short circuit. As a final test, the by-pass switch can be closed to provide full power to the appliance being tested (if tester has not indicated a short circuit).

NOTE: This tester will work with simple electromechanical appliances like toasters, blenders, irons, washers, dryers, and dishwashers. Electronically controlled devices cannot be accurately tested with this unit.

DO NOT TOUCH the tips of the probes or the metal fuse-tester contacts, as voltage is present. The tester is a useful tool for someone who has sufficient troubleshooting electrical know-how.

around. One ploy is to offer an absolute lifetime guarantee for fixing or replacing free of charge some kitchen gadget. Once the device breaks down, you find that you can mail it back to the manufacturer, enclosing $9.99 for postage and handling. The device itself costs $6.59.

Other items are guaranteed "for life." The "lifetime" of such an object obviously comes to a sudden end when the object breaks. The guarantee then no longer applies.

Then there is the watch guaranteed "waterproof as long as crown and crystal remain intact. . . ." If water enters the watch, it is an indication that the crown and crystal are *not* intact, thus voiding the guarantee.

● Some people may not care to repair their belongings themselves, preferring to have a shop or service organization do it. Even in such cases, however, one would want to have things last as long, and be as trouble-free, as possible. Minimal maintenance would consist of reading the instruction manual carefully, then carefully following instructions and respecting the familiarity of the manufacturer with his own product. One should never push a device or appliance beyond its limits (overfilling the clothes washer or putting too much detergent in the dishwasher). Respect the product and don't try to use it for purposes for which it was never intended (like mixing paint with an electric kitchen blender).

Annual and consistent minimal maintenance would also include keeping the object clean and dust-free; polishing metal parts and waxing or oiling wooden parts; removing lime deposits that build up in tea kettles, coffee percolators or other water-using appliances with one of the many special compounds designed for this purpose; and removing batteries from objects to be stored. All products with moving parts also need periodic lubrication.

There are many lubricants, with pros and cons for all of them. Our listing, while partial, will cover the majority of applications:

● Petroleum jelly (Vaseline) is generally used for plastic parts, plumbing fixtures/fittings, and anywhere else where water-sealing, heavy enough to stay in place, seems required. Although older repair manuals may still advocate its use, we find that it does not combine well with rubber, neoprene or nylon parts, and may even have a corrosive effect. We recommend silicone cream that comes in a tube. It is longer-lasting, totally inert and slipperier than petroleum jelly.

● Most sewing-machine oils eventually evaporate, leaving a residue that builds up and actually defeats the whole lubrication process. We suggest instead a good precision oil. Lightest-viscosity machine oils come in needle-spout oilers and are most easily found in hobby shops. These would be applicable to sewing machines and other general oiling purposes. In heavier viscosities, machine oils are available as special oils for use in clocks and other places where oil needs to stay in place. Whether synthetic or natural, clock oils are more expensive than general lubricants.

● You should never use any of these oils on electrically heated devices, which are sensitive to moisture and liquids. We recommend the use of graphite or silicone dry lubricant for this purpose.

Aside from a machine failing to perform, the most visible and common damage to household belongings is breakage. Since the things in our homes are made of an ever-increasing number of natural or synthetic materials, we list some basic glues and adhesive agents. Things to be glued can be divided most simply into porous and nonporous materials. Glass, most metal, formica countertops, mirrors and many plastics are smooth and nonporous. Wood, paper, fractured ceramic bowls and plates, fabrics, cork and many plastic foams are highly porous.

● Glues based on animal hides can be used for joints, especially in wood and other porous materials where joints may need to be opened again. The glue is water-soluble. Antique furniture, predating modern synthetic glues, used animal-hide glues, and for reasons of consistency furniture restorers still use it.

● White glue is a good household multiuse agent and a shop glue for furniture making and repair, hobby work, paper and leather.

● For waterproof woodwork and furniture joints, plastic resin is recommended.

● For more specialized waterproofing considerations, such as wooden outdoor furniture, boat building and repairs, waterproof resorcinol resin will do the best job.

● Contact cement is used for laminating veneers or plastic composition sheets to countertops or vertical surfaces.

Two-part epoxy resins will give high strength in the bonding of metal, concrete, glass, tile, brick, china and some plastics.

Glue guns that electrically melt a stick of glue make it possible to do precision gluing in bonding fabrics, plastics, hobby and craft work, as well as in everyday household repairs of a minor nature involving wood and metal. Great care needs to be taken, since droplets spilling from the tip of the gun may be difficult to remove.

Mastic adhesive is a good bonding agent with high water resistance for the bonding of ceiling and floor tiles and for interior plywood, hardwood paneling and gypsum wallboard to wood, plaster and concrete.

Silicone seal is a powerful bonding agent that can also be used as a sealant because it is self-skinning and forms a flexible solid. Since it is also highly waterproof, it can be used to apply a telephone shower, for instance, directly to the bathtub tiling. When demounting, the telephone shower can be pried carefully from the tile, and the thin film of silicone, now in the form of a rubbery blob, can be peeled from the shower's backing plate and thrown away. The shower may then be re-attached somewhere else, using silicone seal again. It will also work well on pottery, porcelain, glass and metal, and is used in weatherstripping, caulking and aquarium construction.

Finally, "superglues" (the one-drop adhesives) will work on glass, wood, metal, rubber and many other nonporous materials. Since a wide variety exists, check the label for specific applications. However, these glues can permanently bond your fingers (or eyelids) together, which may then have to be surgically separated. Great caution is advised.

Because of the faddish nature of much that appears in the marketplace and the continuing forced obsolesence of good products, there is every reason to maintain what you have. In the next chapter we will discuss both sides of the obsolescence question.

9

Is Your Hi-Fi out of Fashion?
Is Your Mink Coat Obsolete?

Just as rapid technological innovations entail unforeseeable ecological and health hazards, so also can rapid "cultural" innovation produce unforeseen social hazards.

—E. J. Mishan, *Making the World Safe for Pornography*

High-fidelity system and high-fashion fur coats are similiar in many ways. While it may be somewhat difficult to listen to an ocelot coat, and the warmth generated by a solid-state amplifier is negligible, they are alike in how they are acquired, maintained, disposed of and, eventually, replaced by newer versions. Both are status objects. Both are luxury articles. Fur coats as well as hi-fi sets supposedly convey a message to others about the taste, financial position and "consumer expertise" of the owner. Both products are subject to marketing manipulations, and both become victims of true obsolescence.

A case can be made for getting rid of almost any old hi-fi set and trading it in for a newer one. Early vacuum-tube-type sets are becoming difficult to repair. Their trade-in value has diminished to nil. The various components of the set must be kept rigidly separated because of heat generation. Every part of such sets is heavy, bulky and not compatible with newer components. Then there are "hybrids" (a combination of tube and transistorized elements) and, finally, today's solid-state sets.

Over the years a comparable transition has occurred from monophonic to stereophonic. After a short flirtation with ambiophonic sound (a system that used four speakers and provided a delay for the two rear speakers, closely simulating the reverberation time of an auditorium), high fidelity moved to "full-surround sound." At this time that means quadrophonic (four speakers), but even this system is now beginning to move toward linear speakers that will cover a continuous wall strip. One Scandinavian firm is developing a system with one "top-of-the-room" ceiling-mounted speaker, all the walls serving as bass-reverberation units. Sixteen tweeters and midrange speakers, spaced throughout the room, complete the setup.

In sound reproduction, records have moved through changes from 78 rpm to the now universally used 33⅓-rpm system. To this there have more recently been added reel-to-reel tape recorders, four-track cassette decks and eight-track cartridge units. Two gadgets, just released, are large frequency filters with a mind-boggling array of slide controls that enable the user to modulate, tune in or tune out almost any part of an audio message. There are also new audio-oscilloscopes, enabling one to view the *patterns* made by the sound. (A former music critic of the *San Francisco Chronicle* would delight in "viewing" Beethoven for hours, with the speakers turned off.)

Again the wheel has come full circle: Today's "audiophile" will listen to music on a headset not that dissimilar-looking from the earphones on which granddad listened to WOR in 1921.

•

It is just as simple to view one's fur coat as obsolete. Animals go through moulting seasons, so do furs—except that, being cut off from biological processes, a fur coat will only moult once. The price of the fur coat will reflect the rareness of the animal, the animal's size, and how well the individual skins have been matched. (The smaller the animal, the more skins are needed. The more skins there are, the more difficult the matching procedure.) With small, long-haired furs, loss of hair is most rapid.

The relationship of the animal's natural habitat to the climatic conditions in which the fur coat is worn also has a bearing on the longevity of a fur coat. During the summer, the natural moulting season, fur coats are stored in refrigerated vaults. If neglected, the animal's body oils deteriorate (as they will begin to do in any case, once the hide has gone through tanning) and the skin will lose suppleness, crack and separate, all this accompanied by loss of hair. A $12,000 chinchilla coat, unless well cared for, may only last three to six years.

Fur coats, however, are also subject to other constraints. Since they are high-fashion objects, they obsolesce extraordinarily quickly—that is, they become unfashionable. The degree to which they are unfashionable has to do with length (vis-à-vis other apparel), the coat's cut, and even the choice of the particular fur itself. Tibetan lamb and Karakul sheep have been subject to many fashion fluctuations during the last forty years.

To these difficulties should be added the socially unacceptable impact of killing fur-bearing animals and eliminating entire species. While some furriers (such as Kaplan of New York) show remarkable restraint in refusing to deal in furs of wild animals, people continue the killing, and will continue until demand is brought under control. The annual clubbing of baby seals in Canada, after being publicized through the media, has declined because of less demand for seal coats. We advocate synthetic furs. They will last indefinitely, need less maintenance and are less costly. Furthermore, they are easily altered to avoid obsolescence (real fur coats virtually have to be rebuilt), they store well and don't need special climatic provisions. Most important, they don't necessitate the killing of endangered species, and in most cases they are virtually undetectable from the real fur.

To get back to hi-fi. It is a paradox that the most expensive sets (presumably destined for large rooms and private homes) carry an extra control, variously labeled "muting," "contour," etc. These controls permit the set to be played at low volume without significant sacrifice of high and low frequencies. In contrast, inexpensive hi-fi systems (presumably going into student residences, small rooms and apartments) carry no such controls and hence sound best when played "wide open." Like the "power-miser" control on automatic dishwashers, the volume control feature is frequently bypassed by the owner. A lot of people go into hock to buy an expensive and powerful set, then still play it with the volume way up, just as they were forced to do on their earlier, cheaper equipment.

The statistics on deafness in young people in technologically developed countries are staggering. But "power" is a word that still seems to turn people on. Whether it is 427 cubic inches under the hood of a car or a 400-watt hi-fi amplifier, people seem easily persuaded by such claims. Wembley Stadium, near London, manages to get by with amplification of around 50 watts. Unless your living room will seat 300,000 people and has a central, entirely untenanted area more than a mile long, there seems little reason for this "biggest is best" route. One of the finest European high-fidelity systems is quadrophonic, with a 7½-watt output per channel.

What is more to the point is that many of these wattage numbers mean absolutely nothing and are used to mislead the buying public. There are at least four differing ways of evaluating output wattage, yet not one of them has been accepted by manufacturers as a standard way of evaluation. In reality, room-filling volume is determined by speakers being correctly matched to the amplifier. This has little to do with unevaluated huge wattage claims.

The myth that only men are beset by the complexities of high technology is belied by the fact that women are in more constant working relationships with high-technology products, and hence just as exposed to things that don't work.

To acquire things costs money, to continue owning them costs more. Conspicuous consumption costs most of all. Having known several professional photographers who carefully nestle their exotic cameras and equipment in foam-lined aluminum Haliburton cases, we have also been present at their moments of truth. Present international airline security regulations make it impossible to carry any but the smallest of these cases on board. The case must be checked through. At the destination there are several possibilities: The gleaming aluminum case bounces buoyantly off the conveyor belt because, somewhere along the line, it has been emptied of all contents by someone instantly recognizing what the contents would be. The second possibility is that the case will arrive looking like a garbage pail run over by a steamroller. Dallas–Fort Worth, Mirabel and De Gaulle are just some of the many airports where electronic mangling machinery can have this effect.

Finally, and most probable of all—no case. A professional photographer we know may be carrying two Nikons, one Hasselblad and fifteen lenses in his "on location" case, representing about $9000. Under the Geneva Convention the airline is only responsible for a sum *"not to exceed $100."* In several countries additional luggage insurance cannot exceed $300 per piece. The professional photographer's only alternative is to carry private insurance, with premiums going up every time there is a loss, as well as the additional bother of continuously replacing equipment and explaining away the newly acquired gear at Customs points. Photo cases, like gadget bags, act as signals to thieves. They are powerful signals, because they were designed to acquaint everyone with the fact that status objects are being carried.

●

Somewhat timidly we suggest an alternative to in-flight camera storage. By all means have your fitted Haliburton case—it will protect the contents through padding and shell construction—but shove case and contents into some disreputable satchel picked up at Goodwill or the Salvation Army. Before using it, wrench off one lock, then staple the flapping sleeve or tail of an old shirt to the inside of the lid so that three or four inches will dangle out. Now tie up the unlocked side with a length of clothesline (see Figure 106). You are now ready to hand bag and equipment to your friendly airline.

We realize that we just gave this particular game away. No matter—many other ways can be devised to "disguise" your possessions. But isn't this point, although facetiously made, interesting in itself? If you own something valuable, the fact needs to be disguised. This alone may help to remove many people's "pride in possessions."

In fact, all luggage doesn't work. Suitcases are still based on designs dating back to the Victorian age, when porters were plentiful. Today's air traveler will find that porters are frequently few and far between; in violation of FAA regulations there are often no porters available at all in the late evening at such places as Kansas City International Airport, Dallas-Fort Worth and Wash-

106 Exotic camera case

107 Arnold Saul Wasserman carrying a Suit Pac and Week Pac buckled together on one shoulder, and a Day Pac in his hand
Photo J. Ahrend, Los Angeles. Courtesy TAG

108 A woman carrying a Suit Pac and a Day Pac
Photo J. Ahrend, Los Angeles. Courtesy TAG

ington National Airport. In some northern European countries, porters don't exist at all. Nonetheless, we stagger along, carrying immense suitcases and valises, and straining our backs or having a last minute go at double hernia.

A complete rethinking of ways to carry one's possessions and clothing has been long overdue. A designer in Santa Barbara, Arnold S. Wasserman, has developed his TAG Modular Travel System, which is available from the Travel Accessories Group at 2210 Wilshire Boulevard, Santa Monica, California 40403. His system is based on three over-the-shoulder bags that snap together in any combination by means of military speed buckles. Modularity permits "building your own" equipment to fit different travel duration and needs. The system consists of a "Day Pac," which is a personal carryall for business briefcases, overnight clothing

and toiletries, cameras, etc. The "Week Pac" is intended for trips lasting seven to ten days and will carry shirts, blouses, socks, underwear, shoes, bulky sweaters, toiletries and other articles. The "Suit Pac" will carry a jacket, vest, four pairs of trousers and ties in a wrinkle-free condition.

All the TAG Pacs are carry-ons, fit under airline seats or overhead racks, and obviate lengthy waits and mashed luggage at the end of the trip. They come in nylon canvas with leather reinforcements and are lined with backpack cloth.

Wasserman began questioning why so many more people are carrying backpacks these days. A backpack is specifically designed to be carried, thus leaving the hands free. Being soft, it conforms to your body and constitutes a lightweight, rugged weather shell rather than a heavy, squarish, decorated box. Backpackers know that the easiest and

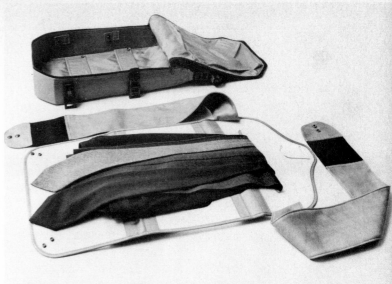

Four small pictures showing the packing procedure of the Suit Pac, and one small picture showing the packing of a Week Pac. 109,
Photo J. Ahrend, Los Angeles. Courtesy TAG

110,
Photo J. Ahrend, Los Angeles. Courtesy TAG

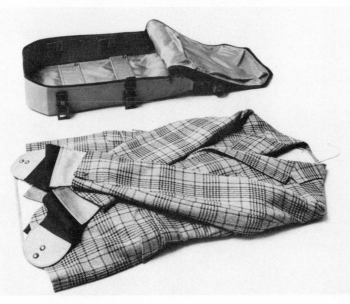

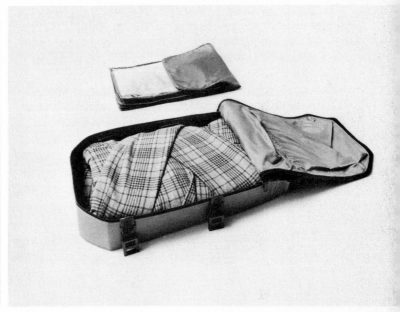

111,
Photo J. Ahrend, Los Angeles. Courtesy TAG

112,
Photo J. Ahrend, Los Angeles. Courtesy TAG

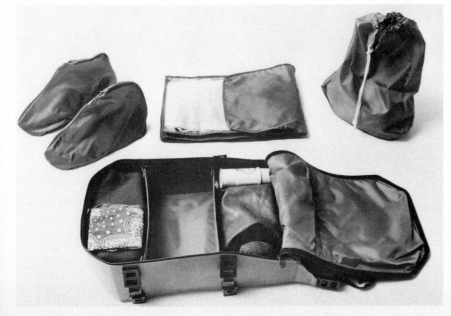

113
Photo J. Ahrend, Los Angeles. Courtesy TAG

safest way to carry a heavy load is strapping it to your back. Unfortunately, backpacks have some problems: Clothes tend to be wrinkled and crushed; a loaded backpack won't fit under an airline seat; and not everyone wants to carry a backpack on a business trip or when wearing a good suit or dress. But the clues for the ideal shape of luggage and how it should be carried were hidden in the present fad for backpacks. A horizontal box can only be carried one way: swinging from the end of your arm like a pendulum and putting excessive strain on shoulder, neck and back muscles. It can unbalance the spinal column and pelvis, causing chronic back ailments. For a good fit to the body a bag should be carried from the shoulder. A shoulder strap permits shifting the weight, rotating the muscle groups used to carry it, and cutting down on fatigue. It also frees the hands. It should be vertical in shape so that weight is distributed along the axis of the body.

Wasserman's modular equipment provides all of these answers in a creative way. Finally the designer feels that he would like to by-pass the typical retail establishment in order to save the customer paying double the prices as well as being served by ill-informed, bored and discontented sales personnel. He goes further in offering to *barter* his products for other ethical goods and services and says, "I don't know if this qualifies as 'alternative

114 The three Pacs in the TAG Modular Travel System. Left to right: Suit Pac, Week Pac and Day Pac
Photo J. Ahrend, Los Angeles. Courtesy TAG

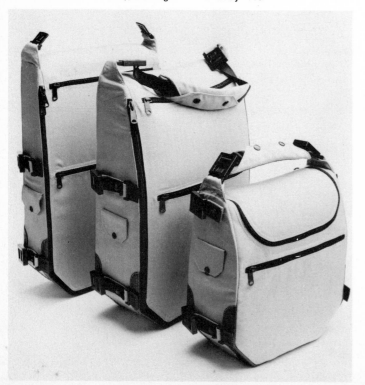

115 The man is carrying a Suit Pac and Week Pac buckled together; the woman is carrying a Day Pac buckled to a Suit Pac.
Photo J. Ahrend, Los Angeles. Courtesy TAG

capitalism' or simply as a revival of the ancient way of exchanging one's own crafted artifacts for other useful products. It may not reform the whole Western mercantile tradition, but the offer already has attracted some interesting responses."

It is a notable contribution by the man who coined the phrase "design entrepreneurship."

To repeat: Owning things is expensive. Because of a general rise in theft and vandalism, insurance premiums also keep rising. They rise more steeply because of ever-climbing replacement costs, caused by materials shortages, higher labor costs and inflation. One-of-a-kind things, collections or antiques are impossible to replace, difficult to repair and prohibitively expensive to insure. With more frequent long-distance moves and life-styles that are generally more nomadic, insurance premiums are climbing more steeply still. For some antiques and works of fine art, insurance premiums are so high that the item is actually fully repurchased every four to five years. Insurance premiums for young people (especially in California, Texas and New York) can amount to nearly one-third of the full price of a particular luxury sports car, which is the most frequently stolen and/or vandalized automobile.

To these purely financial costs must be added the constantly nagging fear and anxiety about fire, destruction, replacement and, last but not least, deterioration or eventual technological obsolescence.

Manufacturers, marketing and advertising (or: the makers, the dealers and the pushers) manipulate us through obsolescence. The connections between fad, fashion and forced obsolescence have been made abundantly clear, but obsolescence operates under many other guises. The late Stewart Alsop fell to musing while hospitalized and wrote:

> It is an old, sound rule of aesthetics that a substance or object to substitute for or imitate a real substance or object is almost always ugly, and almost always less desirable than the real thing. It is a second sound rule of aesthetics that an object with no functional purpose is also almost always ugly. The "simulated-wood estate-wagon-type exterior," besides being fake, is of course an imitation that performs no useful function at all. It is simply intended to give a false impression of money, by imitating, however unconvincingly, the "beach wagons" of the post-Gatsby era, which really were partly constructed of wood.
>
> A person who pays a useful sum of extra money for something that is ugly or useless is by definition a fool: "One who lacks sense; a silly or stupid person."
>
> But I suspect that there are profound conclusions to be drawn . . . about the system of basing the profitability of the distribution system on selling people things like simulated-wood slabs they don't really need, and even about the American Capitalist system, which every year tends more and more to substitute unreal, symbol values for true values.[1]

We will try to break down many of the various ploys that can be subsumed under the heading of obsolescence, and give examples of each. In some rare cases, of course, the obsolescing of a product is unavoidable, and based on good reasons.

Lately the trend toward inexpensive and ultra-compact calculators has speeded up. Only nine years ago a university professor returned from Switzerland having gleefully purchased a simple electronic calculator, nearly pocket-size, for just under $1000. As this chapter is being written, local discount department stores are offering small electronic pocket calculators for $3.95. This is not much more than the price of the batteries, which must be bought separately. By the time you read this book, simple calculators may in fact be giveaway items, free with the purchase of batteries.

Extreme ultramicrominiaturization, a spin-off from the aerospace industry, as well as complete saturation of the market, has brought such prices about. Digital watches, including LED and liquid-crystal displays, have followed much the same pattern. Early models sold for around $800, current models sell for $19.99, and they too will predictably end up at the $3.95 level.

There are a lot of advantages to a watch with one-second-per-month accuracy, no moving parts and an expected life-span surpassing that of the wearer. Such watches are relatively shock-resistant, operate well at temperature extremes, and their tiny batteries only need to be replaced about once a year. (Several watches on the market absorb energy directly from a solar cell and don't need batteries at all.) It can be expected that digital watches of all kinds will eventually replace mechanical watches.

The disadvantages seem more subtle. We tend to inform ourselves about time in relation to elapsed and upcoming events. The statement "It is now exactly 11:15 A.M." is an absurd and meaningless one, unless you have to be at the airport at, say, 12:30. Even if there is nothing else to do on that particular day, your nearly subconscious reaction to such a time announcement is liable to be either, "The morning passed rather swiftly," or "Eleven more hours of freedom left."

The nondigital watch with its open-ended display of all twelve hours and the relationship of the hands to these indices gives us the possibility of making a great many interval judgments when glancing at watch or clock. By contrast, the digital wrist-watch gives us a bald statement of hour, minute and second, and forces us to relate this information to clock-hand positions, which we carry in our head. Thus, small children who have been brought up in a digital-clock environment are usually able to tell time earlier, but are unable to "feel" what it really means.

Both pocket calculators and digital watches are replays of the transistor-radio story of the fifties. Early pocket-size radios sold for over $200; within a few years they were down to the apparently magic sum of $3.95. *These were examples of technological obsolescence, brought about through sophisticated manufacturing methods and integrated circuit design.* The fact that the items are small is relatively unimportant: LED watches tend to be the same size as conventional mechanical ones.

But the drive to smallness (what Bucky Fuller calls "the ephemeralization of the object") *has provided a second way in which things become obsolete.* Shrinking of photo cameras resulted in the Pocket Instamatic 110 format camera some three years ago. Even smaller precision cameras, taking pictures on 16mm-size film, have existed for many years in Germany and Japan. These ultra-miniatures were originally developed for espionage during the war, and were later offered to the public as convenient and easy to carry. Unfortunately they don't work well as cameras. For ultraminiatures, a whole series of wallet-size aluminum sheets have been created by secondary manufacturers, which are meant to plug onto the camera and thus enable the user to hold the camera roughly parallel to the horizon.

Some of the optical drawbacks of the 110 format were examined in Chapter 5. The great virtue of these cameras is that they can be carried easily (thus, as mentioned earlier, leaving the rest of the camera equipment quietly moldering under the bed).

During the spring show of photographic products in Cologne, a third generation of such cameras was unveiled. Now equipped with battery-driven zoom lenses and snap-on close-up devices, the cameras have retained their small front panel, but the rear of the body has grown three to four times. Now they look and weigh like movie cameras held horizontally. Again the wheel has come full circle: Ease and convenience are gone. Larger cameras were obsolesced through the introduction of small ones; now the small ones, with the addition of flash-cube extenders, electronic wink lights, zoom lenses, close-up and wide-angle equipment, cable releases and magnifying hoods, are as bulky and nearly as heavy as the cameras they replaced. Only

the quality of the picture has suffered.

Another device employed to make people discontent with what they now own is *to replace the product with one that works electrically or electronically.* In many cases the original, hand-operated, mechanical product is superior and works far better. Electric traverse curtain mechanisms, using high-technology linear-induction motors to open or shut drapes in your home, the electrically heated doghouse and the electrically operated necktie selector come most readily to mind. Other tools or devices may make more sense, but pose great repair and maintenance problems once they've been juiced up. Electric typewriters put you at the mercy of electric power: There is nothing as dead as an IBM "Selectric" when the electricity goes off.

With respect to typewriters, it is possible to develop portables that operate from a rechargeable battery pack. It is further possible to design a "hybrid" typewriter, one in which mechanical overrides exist so that it can be used manually as well.

Increasing reliance on electric and electronic servomechanisms can have disturbing social consequences. We were trapped at the TWA terminal at Kennedy International Airport during the New York power blackout. At one point an excited gentleman rushed into the terminal and demanded transportation to somewhere—*anywhere.* His bag carried the New York State payroll, which had to be computer-processed in time for Friday. Of course, computers, obviously, were shut down as well as everything else relying on electric power. Automobiles could only drive a limited range (depending on how much gasoline happened to be left in their tanks), since gasoline pumps too are electric. The social consequences of the payroll not being processed would have included the failure of many small "mom and pop" grocery stores and independent service stations, relying on their customers' Friday paychecks to keep going for another week.

If you ever had the experience of staying overnight in a small-town hotel, planning to make an early start for a flight to another city, you may also have suffered the consequences of even a brief power failure, in which the electric alarm clock stopped running for a while as you slept. Such power failures frequently occur during heavy electrical storms in outlying areas. Even if it is your

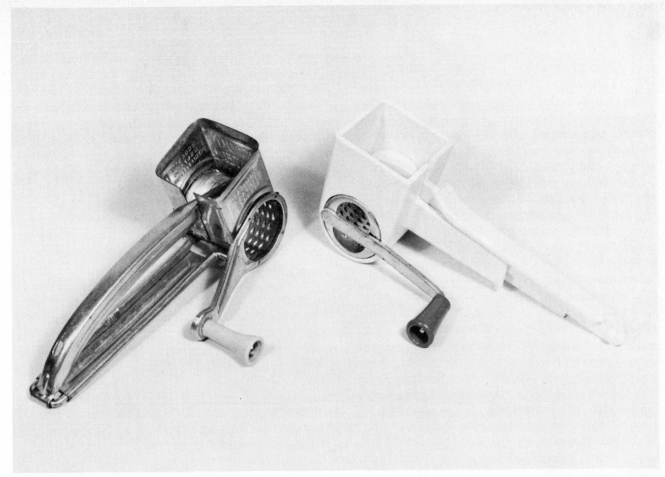

116 Two graters. On the left, the inexpensive, efficient, and nearly indestructible one, which will work left- or right-handed. On the right, the "improved" model, which is right-handed only and, after some months of use, grinds its own plastic coating into the food.

practice to ask to be awakened by the switchboard operator, he or she may in turn rely on an electric clock, equally stopped.

◐ There is a great deal to be said for a small, windup, nonelectric traveling alarm clock, sold everywhere. Here, once again, electrifying it hasn't proven an advantage.

A "gimmicky" redesign of an object frequently makes it work less well. The Mouli grater from France (see Figure 116) is a simple two-part kitchen tool made of tin. It is used for hand-grating cheese, garlic and much else in the kitchen. The design goes back more than half a century. It works well and is used in great kitchens of the world. Recently various grater manufacturers decided to "upgrade" their products by coating them in vinyl. The result is that the vinyl may peel and chip, small vinyl flakes may be mixed into the cheese soufflé, and eventually the whole tool may discolor

and peel so badly that it has to be thrown out. The plastic handle may also break. The original all-tin device, nearly indestructible, can still be found for $1.95. The "improved" version sells for $3.98. We suggest that you shop around for the original.

Marginal improvements are frequently sold to the public, again in an effort to make them discard what they already have. Reaching into the photographic field once more, precision lenses that were superbly designed and operated efficiently were discarded by hundreds of thousands of photographers some years ago for coated lenses. These coated lenses in turn have been succeeded by a third generation, multicoated ones. Optical improvement is extremely slight and, for the amateur photographer, a great deal of extra expense has bought an improvement that frequently cannot even be noticed under a glass.

Photographers especially tend to lose sight of the fact that the great photographers of the immediate past—Paul Strand, Ansel Adams, Henri Cartier-Bresson, Atget and many others—routinely used cameras and equipment that is now available on the used market and in pawn shops for $50 or less.

•

◉ A Japanese manufacturer has marketed a superbly designed "macroclose-up" lens. Their latest multicoated version is identical to their own earlier single-coated offering, which can now be bought very cheaply.

In the frenzied rush to turn people away from what they have, *manufacturers often market things prematurely.* Things are sold that may not have been tested sufficiently. In spite of a rather enviable record otherwise, the Volkswagen company of America apparently erred in this direction with my Volkswagen 411.

About six years ago the Volkswagen 411 squareback was introduced as a theoretical replacement for the less expensive Volkswagen 1600 squareback. The car came with an electronic "brain," which monitored environmental conditions and controlled timing, fuel injection and many other things. Though the car carried excellent warranties, the brain was a sealed unit that could not be examined for malfunctions, only replaced, which made the diagnosis of the brain especially susceptible to the good will and diagnostic ability of the mechanic in any given shop.

My 411 had three different brains installed under warranty after much argument. Some months after the warranty expired, my car, for reasons best known to the electronic brain itself, lurched forward from a standing position while waiting in a freeway lane, accelerated to over thirty miles per hour and slammed into the back of a truck, which also was standing still just ahead.

It took an investigation by the Traffic Division of the Los Angeles Police Department to determine that no driver errors were involved. The automobile's brain apparently could not adjust to the temperature differentials between night and day, which are common at the edge of the Mojave Desert. Both the 411 and the later 412 are no longer made.

"Instant" obsolescence describes products that are manufactured for a brief time period only by fly-by-night outfits eager to cash in on a sudden fad. These operators flood the market with shoddy merchandise, operate from nonexistent addresses, issue meaningless guarantees and crawl back into the woodwork within a few months. If you buy such merchandise, parts and replacement will be un-

available, consequently maintenance and repair an impossibility.

Consider the case of the "Fiskars" scissors. Made in Finland, they are exceptionally well designed and superbly crafted. Ergonomic considerations make them unusually comfortable and also led to a right-handed version (with orange handles) and a red-handled, left-hand version. This design has been ripped off. Inferior versions of this product have been manufactured in Hong Kong that no longer fit the hand as well, use substandard steel, work badly and come only in a right-handed version.

One example of the short-livedness of fad items is the "stereo" slide camera, which was introduced with great fanfare and at high prices in the late fifties. Its introduction was accompanied by special projectors, viewing glasses and "3D" screens. After the public had been media-persuaded to invest millions in all of this, the various lines were discontinued. Today, twenty years later, a few hardened souls and freaked-out experimenters still continue to use the equipment. Others buy it as nostalgia collectibles and store it in cases, unused.

Aesthetic obsolescence is based on image manipulation. The device used is to persuade you to trade in yesterday's clichés for tomorrow's, since in the world of trends "today" never exists. Two strategies are used to aesthetically manipulate the surface of products.

First, an aura of precision and craftsmanship is imparted through the use of chromium and aluminum parts. Some years ago, Jay Doblin, then designer-in-chief at Unimark International, made an experiment about mass-marketing procedures. He disassembled an inexpensive 126-format cartridge camera. Under its shroud of chrome and fake black leatherette, he found a grand total of around ninety parts. Of these, only thirty-five constituted the working mechanism of the camera and were essential to the picture-taking process; *nearly two-thirds of the parts were there just to impress and sell.*

On a grander scale, this superficial attempt to present a falsely impressive front can be found on most control panels of washers, dryers and stoves, as well as on the dashboards of cars, private airplanes, inboard motorboats and snowmobiles.

The visual vocabulary of the cosmetician/designer is so bankrupt that the only other clearly discernible direction is the "military look." This has appeared, over the last three years, on transistor radios and some portable TV sets. The idea seems to be that if you can't have watchmaker's precision, then try for "honest and reliable ruggedness." Since most of these radios are constructed of thin-shell plastic shrouds, painting them olive drab doesn't really lend strength or dependability. All of these radios and TVs have come to look like debris from an army-surplus sale after some future world war.

The real direction functionally and, by derivation, aesthetically should be toward *less* visibility. A well-designed environment or object is one that never intrudes. Thus, a Scandinavian hi-fi system, still in the prototypal stage, is being designed to be located in a closet, with only a remote-control unit (the size of a pack of cigarettes) ever visible in the room. A series of ultrasonic signals control speaker balances, bass and treble, volume, station selection and switching from FM to phonograph or tape modes. The marketing staff of this corporation has decided not to market the set in North America. Their research seems to support the belief that Americans would be unwilling to spend a large amount of money just to hear the *sound*. We crave idiot lights, panels, VU meters, knobs, switches, sliding controls, buttons and toggle switches, with which to stun our friends.

The marketing boys also try to jolly you out of your present possessions *by suggesting that the new product is safer.* Frequently these "safety improvements" only make things seemingly safer, but actually increase chances of injury.

For example, ten-inch circular table saws normally expose the blade. The craftsman, realizing that no spinning saw blade will ever be safe, accordingly takes great care. Since both blade and piece of wood being cut are visible at all times, there is a continuous feedback between hand, eye and brain, and safe work is possible for a skilled person.

Recently, however, a plastic shield has been added as a "safety improvement," complexly articulated to stay parallel to the table surface as wood is pushed through. The shield obscures the work, gets in the way and makes subtle feedback between machine and operator impossible. It also prevents you from using the full potential of the machine in making certain cuts. Unless it is con-

stantly maintained, adjusted and lubricated, it may prevent the movement of the blade through the wood, and lead to more serious accidents. This debatable, minor feature has increased the price of the machine by a hefty sum. While a safety guard is a good idea for the novice, it need not be as large, bulky, or difficult to use as the one described. A simpler mechanism could be designed.

Considerations of safety are routinely used as a protest for protectionism. The Citroën 2CV discussed could no longer be imported into the United States when it did not meet the new safety and anti-pollution standards that were enacted in 1970. (Other Citroën models are still imported.) The result is that the fifteen- and twenty-year-old 2CVs that were already in the country now fetch collector's prices and are gleefully driven by their new owners, while newer, safer and less expensive 2CVs are barred. A host of small low-horse-power and energy-saving European cars are also barred because they do not conform in such matters as bumper height, shape of headlight enclosure, etc. The one real hazard of many of these cars —lack of impact resistance in collisions—might be remedied by licensing them for restricted, e.g. in-city, use only.

"Convenience," "ease of handling" and "customer comfort" are the labels under which more attempts are made to have you give up what works now and "trade up." Some of the conveniences are minor indeed, and many never get used at all.

Prospective grandparents constitute one of the great sucker markets. Hundreds of gadgets and devices are marketed each year to help the "little mother" cope. The most baroque of these are furniture pieces that combine toilet, dining table and reclining seat with a stroller, car bed or buggy. Not since the historical wrestling match of Laocoön and his sons with the serpents of Athene have human beings been subjected to the contortions necessary to erect one of these "labor-saving" devices. Cheaply made of chromed steel, with surfaces of pink or blue marbleized plastic and upholstered in the best sweat-producing leatherette, these surrealistic tinker-toys can be a constant source of frustration for parents, and produce bruises and scrapes for baby.

◐ It is best to have a separate high chair, a folding stroller and an individual toilet seat for the child. These and other baby devices can usually be bought at garage sales or be handed down by friends and neighbors. The tricky, collapsible horrors might just as well stay at grandma's house.

For the home craftsman there is a combination power table saw/drill press/lathe/sander/grinding wheel available under various brand names. This all-in-one tool sounds attractive indeed. It would take up less floor space and require less initial cost outlay than would the whole machine array. But this "one-tool-convenience" in practice means dismantling entire parts of it (usually in the middle of some other work) and in effect rebuilding it into another configuration. Hence the convenience is only skin-deep.

Another "convenience" is the plastic garbage bag, sold in every supermarket. While plastic bags are lighter in weight and thus easier to carry than metal or plastic garbage pails, the pails have been forbidden, in many communities allegedly to reduce noise. People are required to use plastic bags, and drag six or eight of these onto their front drive every week. Can the real reason for this be that the bags do *not* biodegrade and also prevent the contents from breaking down? That they work as a more satisfactory, stable landfill for speculative builders and municipalities? In other words, are you paying for plastic garbage bags which you have been sold as "convenient," and which your municipality requires for "noise abatement," just to provide a foundation for a subdivision of tract homes?

"Planned Obsolescence," where the designer and manufacturer collaborate in devising a product that will wear out within a given time-span, has received much serious attention. On the individual level, it drains away cash; on a global scale, it fritters away irreplaceable resources and energy. Most automobiles require an inordinate amount of maintenance, parts and servicing after about five years, while one European car sold here boasts an eleven-year life expectancy, and a German car maker expects its buggies to cruise on merrily for 100,000 miles and more on U.S. roads.

But nowadays both real and specious technological innovations, wearing the masks of convenience, safety or what-have-you, are increasingly used to plan obsolescing products.

Having given descriptions of the various guises under which obsolescence operates, and also provided examples for these, we would like to present a partial checklist. It may be helpful in making a purchasing decision, but we suggest that you always use it together with our other criteria evaluations (given in previous chapters).

1 *Technological Obsolescence:* Will the item you are considering buying be superseded soon by more sophisticated manufacturing methods?

2 *Size Obsolescence:* Does a smaller, more compact version of what you now have really provide better results?

3 *Powered Obsolescence:* Does it really make sense to have the device or tool steered electronically or electrified?

4 *Additive Obsolescence:* Does the gimmick that the manufacturer has slapped on the newer version improve it at all?

5 *Marginally Improved Obsolescence:* Does the tiny improvement warrant your investing in the newer model?

6 *Constrained Obsolescence:* Has it been tested and proven itself, and have all the "bugs" been ironed out?

7 *Instant Obsolescence:* Is the product only a poor and shoddy copy of the original? Is it a passing fad?

8 *Aesthetic Obsolescence:* Are you buying a trendy exterior, or a phony exterior image of quality?

9 *"Protective" Obsolescence:* Is it just a sales game, or is it really safer?

10 *"Easy" Obsolescence:* Will it really be more convenient, easier to operate, and are other sacrifices implied worthwhile?

It is a sad commentary on the state of two important fields of work that one can always get designers to talk about cooking with great gusto, but chefs rarely talk about design. Cooking utensils, like simple hand tools, have evolved over thousands of years (Figure 117). Hence pans, cooking pots, hand stirrers, whisks, taco presses, pasta makers, sausage stuffers, espresso makers, cake tins, meat cleavers, fowl shears, filleting knives, bowls and other cooking implements all possess strong authentic form, and function superbly well. Only since industry began hiring male designers to make housework and cooking by the "little woman"

117 "Timeless" cooking utensils in wood and tin

118 Detail of weighing scale. The legend "Not Legal for Trade" appears on most kitchen and bathroom scales, an indication of their reliability.

more time-saving, labor-saving, easier, convenient, etc., have really badly designed kitchen utensils entered the market.

Design parameters for most things used in the kitchen or at table now start with the assumption that it must be able to go into the dishwasher. This naturally excludes wooden spoons and stirrers, Greek and French ceramic casseroles, Mexican simmering pots, basting feathers, to name but a few. In general, all early cooking utensils (through about 1930) work, and work exceedingly well. Sometimes they show incredible sophistication in terms of cooking procedures and possibilities of use. Examples that come readily to mind are an Arab couscous pot, a Chinese wok or a shabu-shabu maker from Japan.

Hunting weapons and small arms also show directness and simplicity of design without cosmetic influences. From Kentucky squirrel rifles of the late eighteenth century to the Colt revolver of the nineteenth, there is a straight-line development of aptness for purpose. The Bowie knife of frontier days still makes one of the most satisfactory hunting knives in the world, and is still being manufactured. Eskimos and Indians, with eminent prac-

ticality, adopted the Winchester repeater more readily than anything else provided by white mainstream culture.

Olympic small-bore competition free-style rifles and pistols are superb examples of ergonomic design without any added frills. Competition archery bows have undergone some changes with the introduction of wood laminates and fiber glass, but are still elegantly designed to work well. In the whole area of competition and endurance sports equipment, design tends to be apt, and to provide a good fit between man, activity and environment. This would include javelins, the discus, fiber-glass vaulting poles, ski bindings and boots, especially cross-country skis, rock-climbing equipment, backpacks, tents and sleeping bags, and much else. (See Figures 119 and 120.)

In cooking, hunting and competition sports, obsolescence is caused invariably by the introduction of new materials: in cooking, the change of pans to Teflon-coated inner surfaces; in hunting, the introduction of the compound bow in fiber glass

119 Expedition tent "Morning Glory," by North Face. Designers can look to anthropological data for fitting and dynamic solutions. The famous "black tents" of the nomadic Tuaregs of Chad and Niger inspired this configuration.
Courtesy North Face

120 Expedition tent "Oval Intention," by North Face. At other times designers will look to the latest research in a new geometry for new concepts. This ultra-lightweight tent weighs only 8 pounds, 8 ounces, and is based on Bucky Fuller's geodesics.
Courtesy North Face

or the fiber-glass fishing rod; in sports, the monocoque construction of the sailplane. These all represent authentic major steps forward, *without* "forced obsolescence."

Parameters of obsolescence are imposed from without. They are systemic and can't be fought successfully. Even legislation cannot help completely, since inventions and technological discoveries will continue to be made. The other dimensions of obsolescence tend to be obscure and hard to define, and thus even more difficult to prove. Nonetheless, obsolescence can be coped with, through understanding and recognition of what the process is all about.

10

Can You Participate in a Fire Engine?

The existential response to a successful design, creation, discovery, or invention in science or in engineering, can range from calm satisfaction to absolute rapture.

—Samuel C. Florman, *The Existential Pleasures of Engineering*

Fire engines in Wales look no different from most fire engines in this country. Yet there is a haunting feeling of disorientation when you see the first one, coming down some high street in some Welsh village. It takes a little while to get it all sorted out. It's red, it has a big ladder on top . . . That's it! The ladder! Why do fire engines in Wales, or in small villages in this country, or for that matter anywhere else, carry a six-story ladder when the tallest building is only two stories high?

Once this insight is gained, one can begin analyzing the other components of a fire engine, only to find that the whole thing doesn't seem to make much sense for the place it's in. Fire engines are highly specialized tools—in fact, overspecialized in many cases. But that's the way fire engines come. It makes them extraordinarily expensive to buy, and their complexity makes them enormously difficult to learn and understand. Some of their features can never be tried out in certain locations, because the environment lacks the scope and complexity that the fire engine was designed to deal with. Lest the example of Wales be thought too recondite, houses in many suburban communities all across the country never reach above a second floor, yet their fire engines· can reach fifteen, through an intricate hydraulic and servo-assisted ladder system.

Small rural communities often cannot afford to own enough fire-fighting equipment to effectively serve the region. Some small islands, only a few miles from the mainland, must be written off entirely as far as fire engines are concerned. There isn't enough revenue to buy one, not nearly enough to maintain one, and not many people who can learn how to use one. This holds true of islands off Vancouver (British Columbia), along Canada's Atlantic provinces and the coastal regions off the northeastern United States, the Hebrides and other small islands around Ireland, Scotland and England.

Now consider an ambulance. Another overly complicated vehicle. Because of its many specialty devices, it again takes unusually long to "learn" an ambulance. In the Midwest and South of the United States, undertakers frequently ran ambulance services, until recent legislation which has attempted to eliminate this kind of double jeopardy for the patient. Again, there are many places that ambulances can't go to, because of lack of money or trained personnel. Various islands and sparsely populated areas are often unable to afford this service.

Overspecialization of ambulances frequently covers minor or imaginary needs. In those areas on the Continent where the majority of all accidents

Do-it-yourself emergency service vehicle, based on the Morris Mini-Van (side view). Designed (as are all subsequent drawings) by Alojzi Piatkowski, a postgraduate student at Manchester Polytechnic. 121,

AMBULANCE

occur during rock-climbing and mountaineering sports, ambulances are required to carry a dinghy. Often these ambulances are many miles away from any body of water, but that doesn't matter—that's the way ambulances come.

Anywhere, if there is even the smallest of fires, but someone is seriously hurt, the presence of minimally two vehicles is required: an ambulance and a fire engine. The concept of a nonspecialized general rescue vehicle has never been sufficiently explored. When major disasters hit rural populations, the army is brought in, or the National Guard, and massive help is given. But what about minor disasters? Here a variety of rescue vehicles are used, quite different from hook and ladder trucks, fire engines, etc. But in terms of specialization, expense and training, they are still not the open-ended tools one would hope for.

There are also specialized areas in which rescue vehicles might operate, which have little or nothing to do with routine fire fighting or routine paramedical services. Swimmers and boaters could use rescue vehicles tailored to their needs, as could climbers and mountaineers. There are the specific needs of spelunkers and hikers. Skin divers

and scuba divers have their own emergency requirements, as do hang-gliding groups, people who fly sailplanes and gliders, and many others. To this could be added the special needs of large inner-city schools, playgrounds and parks, or even small industrial complexes.

With these requirements in mind, and realizing the wasteful, high-technology limitations of ambulances and fire engines, a design-research study was undertaken. It proceeded from the basic assumption that a new type of vehicle could be introduced that would fill a space somewhere between fire engine or ambulance on one hand, or a neighbor's car, commandeered for an emergency, on the other. The work was carried out by Alojzi Piatkowski, a postgraduate student at Manchester Polytechnic.

Alojzi decided to investigate the possibility of starting with *any* undifferentiated vehicle that was cheap and easily available. Such a vehicle could be purchased by small groups acting individually and on a decentralized level. There is even the probability that one such vehicle is already owned by a member of the group, and need not be bought at all.

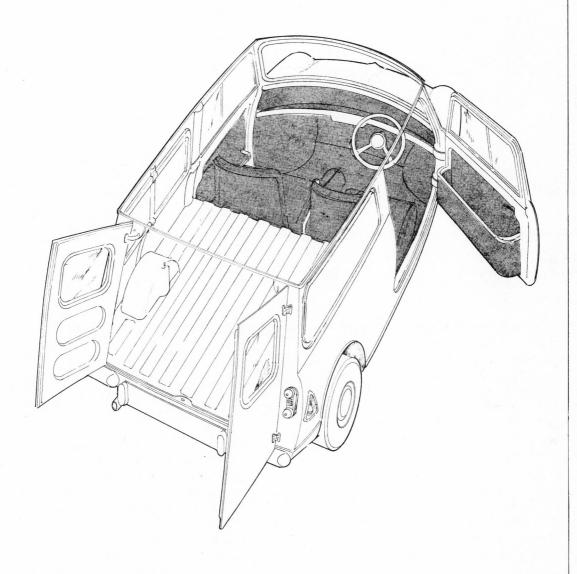

DECENTRALISED EMERGENCY SERVICES
BASIC MODIFICATIONS

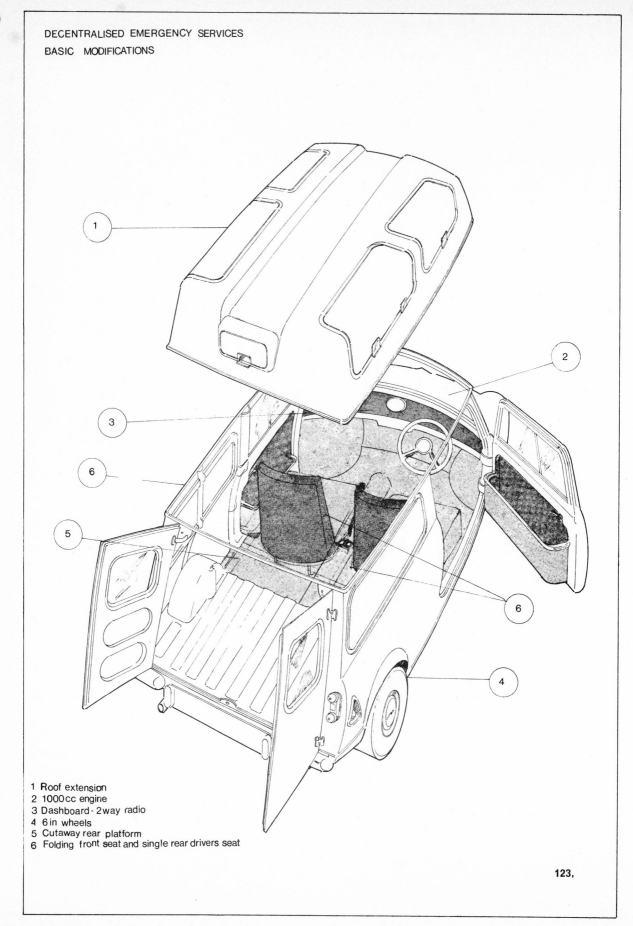

1 Roof extension
2 1000cc engine
3 Dashboard - 2 way radio
4 6 in wheels
5 Cutaway rear platform
6 Folding front seat and single rear drivers seat

123,

DECENTRALISED EMERGENCY SERVICES
AMBULANCE

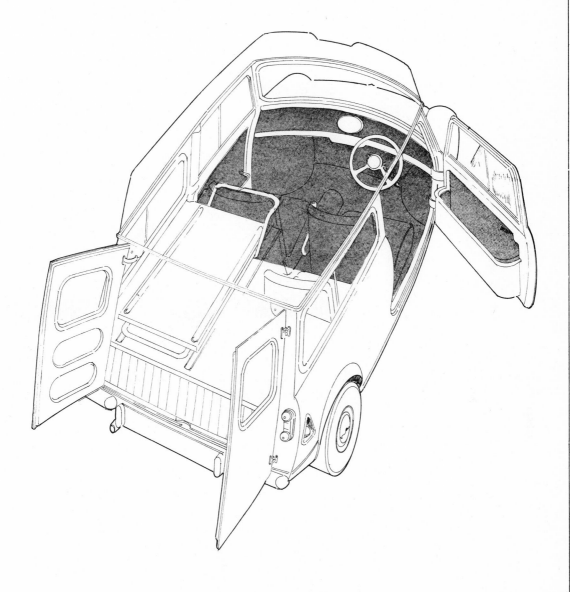

124,

DECENTALISED   EMERGENCY SERVICE
PROPOSED EQUIPMENT LAYOUT

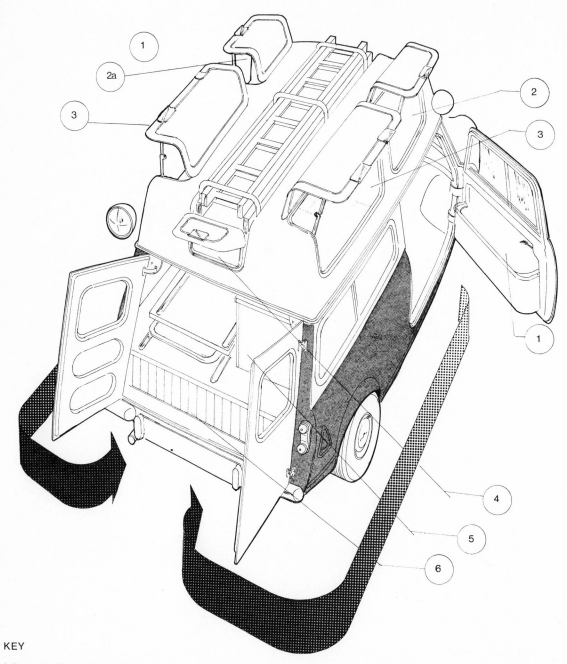

KEY

1 Personal first aid satchel
2 Major first aid store
2bTool store
3 Personal Fire Fighting equipment
4 Stretchers
5 Blankets
6 Portable fire extinguishers

125,

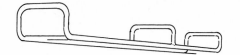

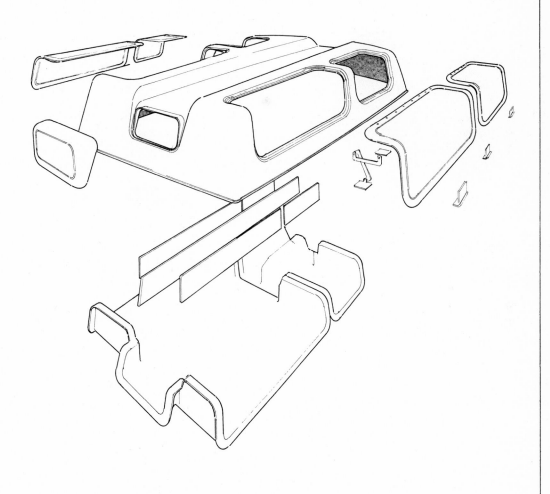

127   First prototype model

Starting with this as a basic premise, it was decided to develop a series of self-help booklets that would make it possible for such groups to partially self-build a vehicle specifically suited to local terrain and local group activities. (See Figures 121 through 127.)

The first step demonstrates how the basic vehicle must be altered in order to make it suitable for general-disaster use. It includes lists of basic supplies and equipment (some of which are strangely absent from the most sophisticated emergency vehicles now in use—for example, several pairs of asbestos gloves, a battery-driven chain saw, battery-driven hacksaw, emergency lighting, etc.).

The second step is to show how vehicles can be divided into several broad areas of emergency disaster aid: agricultural and farm needs, requirements of a large school, marine work, mountain rescue and so on.

The third step is to make this "customizing"

more specific. Here the requirements of a spelunking club would be far different from those of one hundred and fifty people living on an island off the coast of Maine. Nine crofters' cottages in Scotland, or on the west coast of Ireland, have needs far different from, say, two small secondary plastics manufacturers, or a gliding club.

The last and final step is to show precise ways of creating the vehicle and selecting the contents needed by each constituency.

The concept behind the whole plan is that important demands of small groups can most easily be met by each group itself, and that the highly individual needs of each cluster of people can only be met on a decentralized level. The idea is not to eliminate fire engines and ambulances; rather it is an attempt to declassify such vehicles and make them into more accessible, open-ended learning tools for all people.

In the initial phase, a list of the least expensive cars available in various countries was made.

Since size of car also dictates price, only the smallest were chosen. This constituted a trade-off: While it could be argued that a large car could carry two or three injured people, whereas a subcompact might carry only one, it was felt that the prime objective was great maneuverability of the car itself. Subcompacts could enter passageways, footpaths, interior factory or school corridors, pedestrian malls and other narrow areas. While the vehicle might be adapted to carry one man on a stretcher, its primary function was as a mobile first-aid pad and emergency aid center.

It was furthermore decided to assume that the vehicle would still be used by its real owner to transport him and one other person around under normal conditions. Therefore, rescue provisions could not interfere with the primary use for which the owner had originally acquired the car.

The list of vehicles include: the Citroën 2CV and Renault 5 (France), the Volkswagen "Beetle" and "Rabbit"/"Golf" (Germany), the Fiat 126 and 127 (Italy), the Honda "Civic" (Japan), the basic Moskvitch (USSR), and in Great Britain the Austin "Mini-Van."

Since Alojzi Piatkowski did all of his work in Great Britain, his major research was done on the Austin "Mini-Van." All of our illustrations are therefore of that vehicle. In 1974, when research began, it was possible to buy a used "Mini-Van" in Great Britain for the equivalent of $120. Hence it was felt that even groups that did not include at least one person who already owned such a vehicle would be able to acquire one for this specific purpose. The self-help booklets that emerged from this research, however, illustrate step-by-step procedures of transforming and adapting any of the various vehicles listed above.

Transportation systems based on a decentralized mode of life will also need to employ bicycles and bicycle technology. Bicycles are the most efficient locomotion devices so far developed in terms of energy/performance ratio. In terms of body-weight/speed ratio they are far more efficient than a racing car, a 747 jet or a fighter plane. The man-plus-bicycle combination can even outdo such speed phenomena as the cheetah or hummingbird. The statistics of the energy/performance ratio are given in Figure 128, and much additional data is available.[1]

Due to concern with the environment, pollution factors of the automobile, and recent emphasis on the health benefits of physical exercise (Dr. Paul Dudley White, "aerobics," etc.), bicycles have returned to the consumer market and become desirable and fashionable once again. A contemporary lightweight touring bike, with ten-speed derailleurs gears, is close to an ideal tool.

Because of this new popularity and for other reasons, the Japanese Bicycle Promotion Institute, together with their Ministry of International Trade and Industry, and under the sponsorship of the International Council of Societies of Industrial Design (ICSID), held an international cycle-design competition in 1973. In spite of hundreds of entries from all over the world, no major useful suggestions or important breakthroughs were achieved. (The one exception is discussed below.)

This singular lack of success was repeated one year later when the *Internationales Design Zentrum,* in cooperation with the *Rat für Formgebung* in Berlin, held a second similar competition and exhibition. Many entries were received and many ingenious design features were suggested, but again no real breakthrough was made. All this seems to confirm that the muscle-powered two-wheel bicycle has by now achieved its optimal configuration.

That is not to say it cannot be improved. Since it is already incredibly efficient, further improvements, relating to such devices as elliptical gears and "stepless" transmissions, will be expensive and constitute a fine honing of an already highly developed mechanism.

☯ Two areas that can stand further exploration have to do with how much the bicycle weighs and its transportability when not being ridden or pushed, as well as its method of manufacture.

We could then also ask: Can you participate in a bicycle?

☯ Reinder van Tijen has concerned himself with both of these points. The bicycle he has developed, shown in Figure 129, is a folding bike that can be used both as a means of transportation and for recreation. The explosive growth of modern cities often means that public transport such as trams, subways, commuter trains or buses must be used in conjunction with the bike. Therefore it must be possible to ride the bike to the bus stop, fold it up and take it with you. At the final destination point you would then unfold it and continue your ride.

Many folding bikes now exist and are used in

128 Man-on-bicycle performance chart

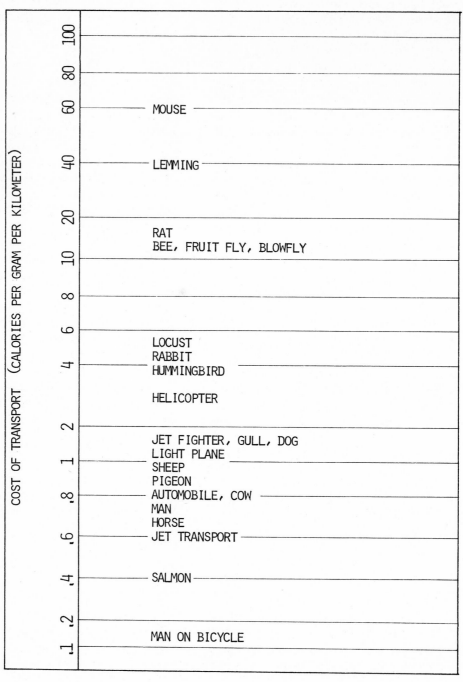

The graph is based on information originally collected by Vance A. Tucker of Duke University. The efficiency of a man on a bicycle (about .15 calorie per gram per kilometer) is even greater than that of an unaided walking man (about .75 calorie per gram per kilometer).

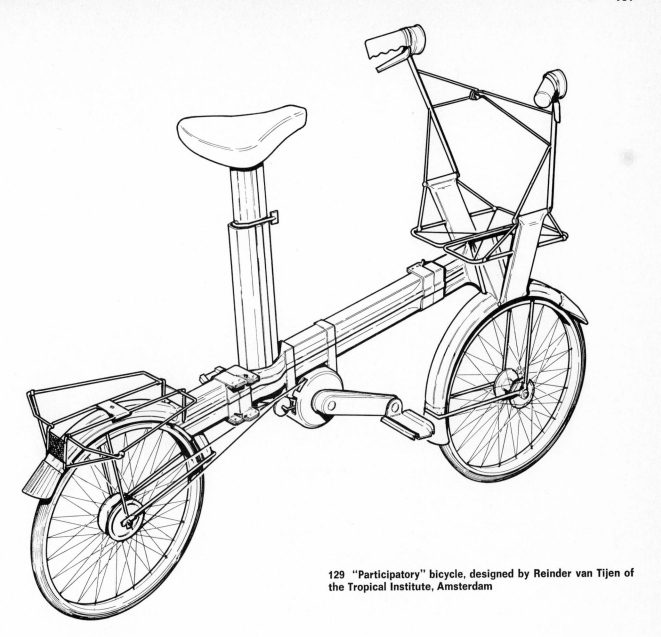

129 "Participatory" bicycle, designed by Reinder van Tijen of the Tropical Institute, Amsterdam

Europe in precisely this manner. But van Tijen's central concern was with manufacturing methods. His inspiration was an old kerosene-powered railroad lantern made of tin. In this device he found a reasonably sophisticated product, with a few crucial tolerances, yet such lanterns are made in a decentralized way, and with simple tools. We quote from Reinder van Tijen's submission to the International Cycle Design Competition:

Though cycling is a healthy occupation, the production of bicycles is not. Large-scale production of an article designed for craftsman-

ship is difficult. Even though the so-called smaller bicycle factories are already producing a turn-over of a quarter of a million, and in spite of high capital investment, the results are meager. The produced bicycles perform well, but they are soon attacked by rust and loosening parts. The price is relatively high. One can buy many good-quality folding chairs for the price of one low-quality folding bike.

Therefore our second aim should be the designing of a bike which is simpler and more easily produced. I designed a bicycle with very low tooling and machine costs, uniform-

130 The Bickerton bicycle

131 The Bickerton, folded

132 The Bickerton, showing its travel container being used as a carry-all. This protective bag was designed by Mitch Fry, a student at Manchester Polytechnic.

ity in the materials used and very simple assembling. Almost every step of the production can be made by manpower.

The fact that it involves low initial capital investment should make it possible for really small bicycle factories all over the world to start production. As there is a desperate need for smaller social systems, this may turn out to be the most important feature of the design.

The concept is that the industry can be redesigned to serve all people, no matter whether they are involved as producer or user or only living together.

Is this "soft" attitude towards technology impractical, not to be realized, unscientific?

The tin bicycle will be:

1 Cheap and durable.

2 Compact and foldable to travel with you in any other kind of transport.

3 Marking a more effective use of labor, materials and energy in the course of its production

This whole product is intended as an example of the way soft technology can be introduced in today's society in an effective way.[2]

Van Tijen's folding bicycle, designed for decentralized methods of production, is still comparatively heavy, which makes it difficult (after it has been folded) to lift it aboard a bus or tram, or check it through with an airline. *Non*folding touring bikes weigh between thirty and thirty-eight pounds. Racing and competition bikes, also nonfolding, tip the scales at twenty to twenty-five pounds. Conventional folding bikes are a heavy thirty-five to forty-six pounds, which defeats their main purpose—portability.

Harry Bickerton of Welwyn, Hartfordshire, England, is a former design engineer for Rolls-Royce, with a great interest in bicycles. Over the last few years Bickerton has developed what seems at present to be the ultimate in a *lightweight folding* bicycle. (See Figures 130, 131 and 132.) Utilizing aluminum and magnesium alloys, Bickerton's folding bike weighs out at eighteen pounds, nine ounces; or at twenty-two pounds with a Sturmey-Archer three-speed gear. It is no heavier than non-folding touring bikes or many nonfolding racing bicycles, a considerable feat. The folding mechanism of the "Bickerton" has also been simplified so that, with practice, folding or unfolding will take less than sixty seconds. Finally, the "Bickerton" also folds down to a smaller package than all other folding bikes; it can go into a standard two-suiter suitcase.

While in Britain, we used this bicycle in conjunction with air transport to the Continent. This was the point where part of the system proved itself unworkable. A folding bicycle as ultralight-weight as the "Bickerton" is destined to be squashed flat or pulverized by most luggage-handling systems. The answer can't just be to stick the bicycle into a suitcase for protection. For if that is done, what is one to do with the suitcase at the other end of the journey? Upon arriving in Germany, unfolding the bike to cycle into, say, Düsseldorf, what do you do with a larger, heavy, stiff container?

With Mitch Fry, a graduate student, we developed a series of lightweight bags that would hold the bike and, through its padded, quilted, foam construction, protect the bicycle completely while in transit. At the other end, the bag would refold and become an outsize carrying sack, hanging between the handlebars and able to carry all manner of gear—or, back at home, a full week's marketing. The bag itself weighs three pounds, eight ounces. This means that bike and bag together still weigh less than most nonfolding touring bikes. And finally an unusual spin-off: The bag (when not hanging from the bicycle) can also become a warm sleeping bag for a small child.

Bickerton seems to be forced into a "small is beautiful" mode of production by sheer economics: His miniplant turns out four or five bicycles a week. This low production rate makes the "Bickerton" somewhat expensive. As interest in the bike in-creases, so, presumably, will orders. Even though Harry Bickerton seems determined to continue production in individually small facilities, he would gladly develop other factories, still employing just a few people but scattered in many different places. Through bulk buying and increasingly efficient production, the "Bickerton" would become much cheaper, eventually reaching the inventor's target price of slightly under $80. So we can say: *"Wait a little longer and the price will come down."*

This same statement, "Wait a little longer and the prices will come down," seems to hold equally true with current trends in consumer electronics. The corollary is: *"If you want to be the first on your block to own a new device, you'll pay through the nose for the privilege."* In the previous chapter, we saw the first owner of a pocket electronic calculator paying 220 times more than current prices for a comparable product. The time lag in that case was seven years, but the downward price curve has been accelerating. It can be predicted that the price drop between initial introduction of a consumer electronics device and the final discount price reflecting market-saturation figures will take less and less time in the future.

The shrinking size of microminiaturized electronic parts will modify the sort of kit-building we discussed in Chapters 3 and 4, in ways not easily foreseeable. And it will have equally far-reaching but as yet unpredictable effects on fixing things yourself. As parts get smaller, it becomes less likely that consumers can readily work with them. A 2mm-square, integrated circuit board, containing over two hundred transistors, thirty diodes and a hundred and fifty resistors, can no longer be wired in, or repaired, by the owner. It is unlikely, in fact, that *anyone* will attempt to repair such a part; instead it will be replaced. Conversely a component that consists entirely of nonmoving, nonwearable parts will probably carry an indefinite guarantee.

But the Heathkit catalog isn't getting any thinner. Do-it-yourself electronic kits will increasingly rely on plug-in subassemblies; and new ultrasmall soldering systems, tools and connectors will be developed. This may lessen the interest of some fix-it-yourself people, since they will find themselves using incomprehensible subassemblies, but it will increase interest in kits on the part of others, because new areas will become intriguing and available on the kit market.

Without advocating it, we can see logical lines from the mechanical wristwatch–digital wristwatch concept; combined with the "weathercube" (a three-inch cube radio permanently tuned to the national weather forecast); the WWV continuous time announcer; and a similar compact device permanently tuned to a twenty-four-hour news broadcast. The resulting wrist-"watch" announcing module, roughly the size of a watch, would no longer have hands or digital read-outs. Instead it would, on button "command," relay verbal information from a radio channel.

Such new devices, coupled with the incredible growth of citizens-band radios during the last year, the further growth of remote-controlled model airplanes and boats and the general demand for new allocation of radio frequencies on all levels, will lead to an eventual total saturation of the radio spectrum. This dense packing of frequencies will make fine tuning exceptionally difficult and lead to much greater use of closed-circuit cable transmission.

With more households served by cable, and with greater differentiation of interests and needs among individual subscribers, the emergence of hundreds of differing TV channels through multiplexing seems safe to predict. In Great Britain, the BBC has already introduced the CEFAX system in several cities. This TV attachment makes it possible for individual subscribers to select areas of specific interest, and receive specialized information. Through the coupling of telephone cables and multiplex messages, we foresee "responsive" TV services—that is, the possibility to talk back to the telecast originator. The adaptations of this to elections, public-opinion surveys, preferential primaries, education and shopping seem self-explanatory.

Since this is a practical book on how to cope with the technological age *now,* we feel somewhat uncomfortable writing futuristic scenarios. The greater frequency of historical discontinuities makes the job of prophet increasingly hazardous, and we find it difficult to share the visions of the futurists.

Buckminster Fuller still places his faith in technology: Whatever is technologically wrong, more technology will straighten out. Implicit in this philosophy is the notion that technological change can work independently from politics. The recent United Nations "Habitat" Conference on Human Settlements in Vancouver, Canada, has shown the

naïve falseness of this position. The main topic that engaged the conference had to do with supplying clean water to everyone on Earth. But the conference itself broke down over political haggling on the issue of racism.

Hermann Kahn's faith is pinned on a continuing growth of the gross national product. But consumer opposition is making itself felt most in the advanced technological societies of the United States, Canada, Japan, Sweden and Denmark, Great Britain and Germany. While the second report of the Club of Rome paints a slightly more attractive picture than the earlier one, it, too, is predicated on minimal or marginal growth.

The science-fiction fantasy of wall-to-wall 3-D TV, envisioned by Ray Bradbury in *Fahrenheit 451,* does not seem credible to us. Certainly trends exist that back up Bradbury's horrific vision. The perfection of projection TV, coupled with the experimentation in 3-D holographic imagery, seems to point the way. But more recent tendencies would indicate an early reversal of this course. Smaller-screen TV sets are being bought, and less time is being spent watching television. Among the very young, an increasing dissatisfaction with TV is becoming apparent. A greater emphasis on direct participation, outdoor sports activities, backpacking, camping, etc., is making itself felt among young people. Travel seems to be supplanting passive spectator entertainment.

Martin Pawley has written disturbingly about the trend toward a more private future. It is his thesis that all the devices supplied to people by design and technology will eventually drive them into greater states of alienation within family, community and society. He goes to extreme (and sometimes touchingly British) lengths to make his point: Central heating is condemned as a possible element in destroying family life, by making it possible for family members to be in several different rooms during the winter instead of huddling around the "gas fire" when it gets cold. His eventual vision is of the individual cocooned into a private "cell," passively fed through auditory, visual and sensory channels.

Pawley regards all of the trends that lead to this vision as irreversible and unstoppable, but nearly everything that we have said in this volume negates his apocalyptic vision. And so does most of what young people have been doing and saying over the last few years. Since their voices of protest come largely from precisely those societies technologically advanced enough to make Pawley's withdrawn pleasure-womb possible, we feel that the trend has already reversed itself.

We have all been in cinemas when a computer fails on the screen or a high-technology product operates improperly, eliciting from the audience wild applause, cheering and a general sense of glee. And along with our frustration, there is often a minor thrill of pleasure when a computer-handled bill or letter is wrong. Our reliance on and enjoyment of high technology is frequently overstated. This is not to give way to the doom-laden views of Theodore Roszak, Jacques Ellul or the naïve meanderings of Reich's *Greening of America*. Rather it is an attempt to cope with technology in a positive way.

At the School of Design at the California Institute of the Arts, the student body came largely from an upper-middle-class background. With curriculum and staff so loosely structured as to make any underlying system virtually undetectable, classes proliferated madly. Finally there were a hundred and thirty students, demanding more than seventy different courses, not to mention an ever-increasing number of courses offered by the rest of the institute. In the end, there were more courses than students.

We feel this to have been the result of early experience with electronic media, which offer mind-blinding numbers of possible options. Living in an upper-middle-class environment in Los Angeles, with dozens of TV and radio channels available on a twenty-four-hour basis, the pushing of a button would provide most desirable options. The attempt to carry this over into the school made the system unmanageable. It is important to note that restructuring of courses, and the elimination of many of the offerings, eventually took place by student demand.

Obviously, options in the future will continue to multiply. These technological options will enslave you to the degree to which you are unaware of the process, and are willing to let yourself be thus enslaved. The confusing array of choices will have a tendency to fragment life to an even greater extent. But we are not Pawley's passive puppets. In the final analysis, your awareness is increased and things are up to you.

Anthony Wedgwood Benn, one of Britain's foremost thinkers and a minister of the Labour cabinet, said:

> Now, all of a sudden, people have awakened to the fact that science and technology are just the latest expression of power and that those who control them have become the new bosses, exactly as the feudal landlords who owned the land, or the Capitalist pioneers who owned the factories, became the bosses of earlier generations. Ordinary people will not now be satisfied until they have got their hands on this power and have turned it to meet their needs.[3]

Note to the Reader

If you have any personal experience or case histories of successfully coping with our technological age, please tell us about them. Photographs of three-dimensional solutions (no color slides, please!), drawings, sketches, plans and brief descriptions can be sent to us as follows:

Victor Papanek and Jim Hennessey
c/o Pantheon Books
201 East 50th Street
New York, N.Y. 10022
U.S.A.

If we receive enough interesting solutions, ideas and proposals, we shall edit them, collect them and publish them for everybody's information. We shall, of course, give full credit to you should such a book become possible.

Notes
and Bibliography

Notes and Sources

We have not attempted to give every one of the hundreds of newspaper items referred to in the text, but to give all those that cite figures or are of unusual importance.

Chapter 1

1 Moen Products, Ohio, untitled and undated circular, presumably 1973–5.

2 Rudofsky (56). Rudofsky combines insight, wit and a truly civilized mind. All his books can be recommended highly.

3 Musselman (33).

4 Kira (28), (29). The revised edition dispenses with such picture captions as "*simulated* male urine stream." Only nine short years ago, in 1967, even *graphs* of urine dispersion patterns had to be carefully labeled "simulated." Young ladies, fully clad in flowered bathing suits and sitting on the edge of a bathtub, had to have their faces blanked out in the 1967 edition. Now they (or their younger sisters) appear stark naked and smile at the camera. Together the two volumes are a curious example of changing mores in nine years.

5 Papanek (39).

6 *Ibid.*

7 *New Zealand Herald,* November 10, 1976.

8 Papanek and Hennessey (46), p. 123.

9 Rybczynski (59).

Chapter 2

1 Thorstein Veblen, *The Theory of the Leisure Class* (1899; reprint ed., Boston: Houghton Mifflin Co., 1973).

2 Rudofsky (55).

3 *Newsweek,* September 23, 1974 (European edition).

4 Whitt and Wilson (75). A highly technical and excellent book on bikes and biking, full of ergonomic and anthropometric data unavailable elsewhere.

Chapter 3

1 Personal communication from Harry Rhodes to Jim. We've tried to get one of those booklets of World War II vintage, so far unsuccessfully. If you know of one, write us.

2 Advertisement, *Popular Science,* February 1976.

3 Heathkit catalog 1976.

4 *Poems of New China* (Peking: Foreign Language Press, 1969), introduction.

5 Papanek and Hennessey (47), p. 139.

6 Rosenbloom (53).

7 Papanek and Hennessey (46), p. 114.

Chapter 4

1 *Time,* May 12, 1975 (European edition), for all three statements.

2 Karl Marx, *Letters to Ricardo* (Dresden: Dresdener Nachdrucksverlag, 1967).

3 Rothschild (54), especially chap. three.

4 Associated Press News Special, *Kansas City Times,* December 16, 1976.

5 As reported in *Dagens Nyhetter* (Copenhagen), April 10, 1974, and in *Time* (European edition).

6 ABC-TV national network news, May 29, 1976.

7 *Design* (London), October 1975.

8 *Newsweek,* July 29, 1974 (European edition).

9 *Sunday Times* (London), December 16, 1973.

10 Rothschild (54).

11 John Muir, *How to Keep Your Volkswagen Alive: A Manual of Step-by-Step Procedures for the Complete Idiot* (Sante Fe, N.M.: John Muir Publications, 1969; rev. ed. 1974). The best, clearest and most thorough maintenance manual ever. A classic.

12 Schmidt (60).

13 *Time,* May 6, 1974 (European edition).

14 *Newsweek,* July 22, 1974 (Asian edition).

Chapter 5

1 Consolidated Edison Company of New York, public information leaflet.

2 *Ibid.*

3 Robertson (52).

4 Sears, Roebuck and Co., picture caption in fall-winter 1976 catalogue.

5 Robertson (52), especially chap. eight.

6 The *Montreal Gazette,* November 1, 1975, quoting the Washington service of the *Chicago Daily News.*

7 *Ibid.*

8 *Ibid.*

9 *Newsweek,* December 27, 1976 (U.S. edition).

10 Papanek (39).

Chapter 6

1 *Time,* June 9, 1975 (European edition); *Newsweek,* June 23, 1975 (Asian edition). Also *Der Spiegel,* June 1, 1974.

2 Papanek (39).

3 Papanek and Hennessey (46).

4 McLuhan (30); still McLuhan's best contribution.

5 Asmussen (1).

6 King (27).

Chapter 7

1 P.S.: Under the title "One-Armed Gold Mines," the magazine *Sweden Now* has just published the following: "Swedish restaurants used to be places where you eat. They still are, if you have any money left by the time you've made your way past the one-armed bandits lining the entrance hall. Milking the public has become the lifeblood of restaurants, sports clubs and people's parks all over the country. The machines are so profitable, some politicians want to forbid them—while others want to nationalize them!" (*Sweden Now* [Stockholm], vol. 10, no. 2, June 1976).

2 Papanek (39).

3 *Time,* June 2, 1974 (European edition).

4 *Time,* June 17, 1974 (European edition).

5 *Der Stern,* no. 24, 1973.

6 *Design* (London), November 1975.

7 "*Domus*" Magazine (Milan), no. 560, July 1976.

Chapter 8

1 Vale (71); this is one of the best books on "autonomous" housing from across the Atlantic.

Chapter 9

1 *Newsweek,* April 22, 1974 (International edition).

Chapter 10

1 Wilson (77); S. S. Wilson is an engineering professor at Oxford University who has built three-wheeler bikes for developing countries.

2 International Cycle Design Competition, *Report on Prize-Winning Designs* (Tokyo: Koichi Ishida, 1974).

3 Anthony Wedgwood Benn, "Technical Power and People," *Bulletin of the Atomic Scientists,* December 1971.

Bibliography

(1) Asmussen, Christian, and Sørensen, Bent. "Kaffeskoldninger" [Coffee-scalds]. *Ugeskrift for Laeger* [Physician's weekly] (Copenhagen), vol. 135, no. 6, February 5, 1973. Reprints are available from Olaf Møllers Bogtrykkeri, Mariensdalsvej 50, Copenhagen, Denmark.

(2) Baer, Steve. *Dome Cookbook*. Albuquerque, N.M.: Zomeworks, 1968.

(3) ———. *Flow Shot 1*. Albuquerque. N.M.: Zomeworks, 1967.

(4) ———. *Sol Shot 1*. Albuquerque, N.M.: Zomeworks, 1968.

(5) ———. *Zome Primer: Zomohedra Geometry*. Albuquerque, N.M.: Zomeworks, 1970.

(6) Bell, Daniel. *The Cultural Contradictions of Capitalism*. New York: Basic Books, 1976.

(7) Bigsby, C. W. E., ed. *Superculture*. London: Paul Elek, 1975.

(8) Boyle, Godfrey, and Harper, Peter, eds. *Radical Technology*. New York: Pantheon, 1976.

(9) Brand, Stewart, ed. *The Updated Whole Earth Catalog*. New York: Random House, 1972.

(10) ———. *The Whole Earth Epilog*. New York: Random House, 1974.

(11) Callenbach, Ernest. *Ecotopia*. Berkeley, Calif.: Banyan Tree Books, 1975. A speculative novel.

(12) Chandler, Barbara. *Flat Broke: A Guide to Almost-Free Furnishing*. London: Pitman, 1976.

(13) Charney, Len. *Build a Yurt: The Low-Cost Mongolian Round House*. New York: Macmillan, 1974.

(14) Consumer Guide Magazine Editors. *How It Works . . . and How to Fix It*. New York: New American Library, 1974.

(15) Daniels, Farrington. *Direct Use of the Sun's Energy*. New York: Ballantine, 1974.

(16) Dickson, David. *Alternative Technology*. Glasgow: Fontana/Collins, 1974.

(17) Fathy, Hassan. *Architecture for the Poor*. Chicago: University of Chicago Press, 1973.

(18) Florman, Samuel C. *The Existential Pleasures of Engineering*. New York: St. Martin's, 1976.

(19) Goodlad, Sinclair, ed. *Education and Social Action*. London: Allen & Unwin, 1975.

(20) Hennessey, James. See items (46) and (47).

(21) *How Things Work: The Universal Encyclopedia of Machines,* vols. 1 and 2. London: Granada, 1972, 1974.

(22) Kahn, Lloyd, ed. *Domebook One.* Menlo Park, Calif.: Pacific Domes, and New York: Random House, 1970.

(23) ———. *Domebook Two.* Menlo Park, Calif.: Pacific Domes, and New York: Random House, 1971.

(24) ———. *Shelter* [also contains *Domebook Three*]. Menlo Park, Calif.: Pacific Domes, and New York: Random House, 1973.

(25) Keats, John. *The Insolent Chariots.* Philadelphia: Lippincott, 1958.

(26) Keller, Goroslav. *Dizajn.* Zagreb, Yugoslavia: Nišp-a Vjesnik, 1976.

(27) King, Alexander. *Mine Enemy Grows Older.* New York: Simon & Schuster, 1958.

(28) Kira, Alexander. *The Bathroom, Criteria for Design.* New York: Bantam, 1967.

(29) ———. *The Bathroom.* Rev. ed. New York: Viking, 1976.

(30) McLuhan, [Herbert] Marshall. *The Mechanical Bride.* New York: Vanguard Press, 1951.

(31) Mishan, E. J. *Making the World Safe for Pornography and Other Intellectual Fashions.* London: Alcove Press, 1973.

(32) Mother Earth News, ed. *Handbook of Homemade Power.* New York: Bantam, 1974.

(33) Musselman, M. M. *Get a Horse!* Philadelphia: Lippincott, 1950.

(34) Oliver, Smith Hempstone, and Berkebile, Donald H. *Wheels and Wheeling.* Washington: Smithsonian Institution, 1974.

(35) Palm, Göran. *En orättvis betraktelse.* Stockholm: Panbooks, 1966.

(36) ———. *Vad kan man göra?* Stockholm: Panbooks, 1969.

(37) Papanek, Victor. "Areas of Attack for Responsible Design." In *Man-made Futures.* London: Hutchinson, 1974.

(38) ———. *"Big Character" Poster No. 1: Work Chart for Designers.* Charlottenlund, Denmark: Finn Sloth, 1973.

(39) ———. *Design for the Real World: Human Ecology and Social Change.* Stockholm, 1969; New York: Pantheon, 1971.

(40) ———. "Friendship First, Competition Second!" *Casabella* (Milan), December 1974.

(41) ———. "Notes from a Journal 12/II/1972–12/V/1973." *Mobilia* (Snekkersten, Denmark), nos. 219–220, October–November 1973.

(42) ———. "Il Progetto Lolita." *Design* (Bergamo, Italy), vol. 3, 1975.

(43) ———. "Project Batta Kōya." *Industrial Design,* July–August 1975.

(44) ———. "On Resolving Contradictions Between Theory and Practice." *Mobilia* (Snekkersten, Denmark), July–August 1974.

(45) ———. "Socio-Environmental Consequences of Design." In *Health and Industrial Growth* (CIBA Symposium XXII). Holland: Scientific Publishers, 1975.

(46) Papanek, Victor, and Hennessey, James. *Nomadic Furniture.* New York: Pantheon, 1973.

(47) ———. *Nomadic Furniture Two.* New York, Pantheon, 1974.

(48) Pauwels, Louis. *Manifest eines Optimisten.* Berne, Switzerland: Scherz Verlag, 1972.

(49) Pirsig, Robert M. *Zen and the Art of Motorcycle Maintenance.* New York: Morrow, 1974.

(50) Princeton Publishers. *Spectrum: An Alternate Technology Equipment Primer.* Milaca, Minn.: Alternative Sources of Energy, 1975.

(51) Rivers, Patrick. *The Survivalists.* London: Eyre Methuen, 1975.

(52) Robertson, Andrew. *The Lessons of Failure.* London: Macdonald, 1973.

(53) Rosenbloom, Joseph. *Kits & Plans.* Willits, Calif.: Oliver Press, 1973.

(54) Rothschild, Emma. *Paradise Lost: The Decline of the Auto-Industrial Age.* London: Allen Lane, 1974.

(55) Rudofsky, Bernard. *Behind the Picture Window.* New York: Oxford University Press, 1955.

(56) Rudofsky, Bernard. *The Kimono Mind.* New York: Doubleday, 1965.

(57) ———. *Streets for People.* New York: Doubleday, 1969.

(58) ———. *The Unfashionable Human Body.* New York: Doubleday, 1971.

(59) Rybczynski, Witold, ed. *The Problem Is.* Montreal: Minimum Cost Housing Group, McGill University, 1971.

(60) Schmidt, Tage. "Citroën Design 1980." *Brugskunst og Industriel Design* (Copenhagen), nos. 8/9, October–November 1973.

(61) *The Shaker Millennial Laws.* Pennsylvania, 1797.

(62) Seymour, John, and Seymour, Sally. *Self-Sufficiency.* London: Faber & Faber, 1973.

(63) *Survival Scrapbook 1: Shelter.* Brighton: Unicorn Books, 1972.

(64) *Survival Scrapbook 2: Food.* Brighton: Unicorn Books, 1972.

(65) *Survival Scrapbook 3: Access to Tools.* Brighton: Unicorn Books, 1973.

(66) *Survival Scrapbook 4: Paper Houses.* Caerfyrddin, Wales: Cymru, 1974.

(67) Taylor, Gordon Rattray. *Rethink: A Paraprimitive Solution.* London: Secker & Warburg, 1972.

(68) ———. *How to Avoid the Future.* London: Secker & Warburg, 1975.

(69) Time-Life Books Editors. *How Things Work in Your Home (and What to Do When They Don't).* New York: Time-Life, 1975.

(70) Thring, M. W. *Man, Machines and Tomorrow.* London: Routledge & Kegan Paul, 1973.

(71) Vale, Brenda, and Vale, Robert. *The Autonomous House.* London: Thames & Hudson, 1975.

(72) Varming, Michael. *Byfornyelse: hvordan, hvornår og for hvem?* Copenhagen: Gyldendal, 1972.

(73) Von Scheidt, Jürgen. *Innen-Verschmutzung: Die verborgene Aggression.* Munich: Droemersche Verlagshandlung, 1973.

(74) de Waal, Allan. *Planer, Facader og snit.* Odense, Denmark: Borgens Forlag, 1971.

(75) Whitt, Frank Rowland, and Wilson, David Gordon. *Bicycling Science: Ergonomics and Mechanics.* Cambridge, Mass.: M.I.T. Press, 1974.

(76) Williams, Duncan. *To Be or Not to Be.* London: Davis-Poynter, 1974.

(77) Wilson, S. S. "Bicycle Technology." *Scientific American*, March 1973.
and these magazines, dealing with alternatives:

Alternative Sources of Energy   *The Journal of the New Alchemists*
Appropriate Technology   *The Mother Earth News*
The Co-Evolution Quarterly   *Solar Energy Digest*
Intermediate Technology   *Undercurrents*

as well as the following journals, magazines and newspapers:

Casabella (Italy)   New Zealand *Herald*
Cree (France)   *Newsweek*
Dagens Nyhetter (Stockholm)   *Penthouse*
Design (England)   *La Prensa* (Spain)
Design (Bergamo, Italy)   *Der Spiegel*
Designer (England)   *Start* (Yugoslavia)
Domus (Italy)   *Der Stern*
Expressen (Stockholm)   *Time*
Form (Germany)   *The Times* (London)
Form (Sweden)   *The Times* (New York)
Helsingen Sanomat (Finland)   *Vorm* (Netherlands)
Infordesign (Belgium)   *Werk* (Switzerland)
Ingenørens Ugeblad (Copenhagen)   *Die Zeit* (Germany)
Montreal *Gazette* (Canada)   *Zero* (Switzerland)

About the Contributors

Jim Hennessey is a designer and faculty member at the Rochester Institute of Technology, Department of Environmental Design, where he teaches courses in both industrial and interior design. He received a B.S. in product design from the Illinois Institute of Technology, Institute of Design, and an M.F.A. in design education from the California Institute of the Arts. He has been a Fulbright scholar to Sweden, was selected to participate in the Braun International Design Traveling Exhibition, and has been the recipient of numerous design awards and honors.

Hennessey is an active consumer and "design sense" advocate, and his consultant work (the development of specialized electromechanical devices based on human need) is indicative of his involvement. Together with Victor Papanek he is coauthor of both *Nomadic Furniture* and *Nomadic Furniture 2.* Professor Hennessey, his wife, Sara, and their three children live in Rochester, New York.

Victor Papanek works as designer, anthropologist, writer, teacher and filmmaker. Educated at Cooper Union, M.I.T., and with Frank Lloyd Wright, he has lived and worked in eleven countries, as well as with Navajo Indians and Eskimos. He works as design consultant for the World Health Organization in Geneva, Switzerland, to develop diagnostic kits for primary health care on village levels in the Third World. He designs for UNESCO/UNIDO in design management for developing countries, and has been a senior design consultant for the Volvo A/B automobile firm in Sweden, developing a taxi for the handicapped and creating work-enrichment programs for the workers.

Papanek's book *Design for the Real World* has been translated into twenty-three languages, making it the most widely read book on design. He has also written *Nomadic Furniture* and *Nomadic Furniture 2* (with Jim Hennessey). His *"Big Character" Poster: Work Chart for Designers* was published in Denmark (in English) and has since 1973 been used by designers throughout the world.

After resigning as dean of the School of Design at CalArts, Papanek was a visiting professor for one year at the Royal Academy of Arts in Copenhagen in Denmark, two years at Manchester Polytechnic in England and one year in Canada. He has been elected to a Professional Fellowship of the Society of Industrial Artists and Designers (Great Britain), is a member of Industrielle Designere (Denmark), an Honorary Fellow of the American-Scandinavian

Foundation, and a Diplomat and Fellow of Diseñadores Industriales (Instituto Téchnico Político Nacional) of Mexico. He is presently chairman of design at the Kansas City Art Institute and lives with his wife, who is a designer and weaver, and their six-year-old daughter in Kansas City, Missouri.

Ira Velinsky, who helped with some of the book's designs and illustrations, is a recent graduate of the Department of Environmental Design at the Rochester Institute of Technology. He specializes in interior and product design, and plans to study for his master's degree under a grant at Cornell University.

Mike Whalley, who also helped with illustrations and design ideas, is a postgraduate industrial-design student from Manchester Polytechnic. He is married to a design teacher and lives at Rusholme, England.